MOTORING SHORTS

TONY BOSWORTH

GUINNESS PUBLISHING

ACKNOWLEDGEMENTS

I am grateful to a number of people for pointing me in the right direction when I was researching Motoring Shorts, but I would especially like to thank Nigel Blundell, Associate Editor at the *Daily Star*, and Dave Williams, Motoring Editor at the *Daily Express*, whose help and assistance was invaluable. Thanks, chaps!

Copyright © 1994 Tony Bosworth

The right of Tony Bosworth to be identified as the Author of this Work has been asserted in accordance with the Copyright, Design and Patents Act 1988.

Published in Great Britain by Guinness Publishing Ltd, 33 London Road, Enfield, Middlesex

All rights reserved. No part of this publication may be reproduced, stored in a retrieval system, or transmitted in any form or by any means, electronic, mechanical, photocopying or otherwise without prior permission in writing of the publisher.

Front cover illustration by Frances Button

Design and layout by Kathleen Aldridge

Typeset in Palatino by Ace Filmsetting Ltd, Frome, Somerset
Printed and bound in Great Britain by The Bath Press, Bath, Avon

'Guinness' is a registered trademark of Guinness Publishing Ltd

A catalogue record for this book is available from the British Library

ISBN 0-85112-637-5

ABOUT THE AUTHOR

Our picture shows Tony Bosworth re-enacting one of his early motoring experiences. Careful readers will note that his sunglasses are on upside down, a factor which led to the look of faint surprise, some would say horror, just before the accident. Yes, the hapless Bosworth's first experience with the motor car involved a crash. It happened like this . . .

He was at the wheel of the open-topped car, going as quickly as he could, when all of a sudden he lost control. He tried to brake, he tried to swerve, he tried everything his driving experience had taught him – but it was all too late. He crashed into a tree. He then compounded his mistake with a further unwise decision – he decided to run from the scene of the crime, leaving the crashed car and hot-footing it back home where he rather bizarrely chose to hide in a wardrobe.

In fact, as history records, all of this was to no avail because half an hour later there was a knock at the door. He was dragged outside and questioned. 'Tell me why you took the car. Why didn't you stop when we shouted after you? Why didn't you bring it back in one piece?' All of this went over the head of struggling Bosworth. Mind you, he was only four years old at the time and as he remembered it afterwards, it was something to do with getting his feet caught in the car's pedals.

These were the pedals for pedalling you understand, not the accelerator or brake – this car didn't have those. The pedal car belonged to his best mate, Mike, and the retribution was swift. Bosworth received a clip around the ear and was made to say sorry. The car was towed from the scene, and from then on was never left out overnight on Mike's front lawn.

After that first vivid motoring experience, Tony Bosworth knew one thing: he wanted to be in cars, but hopefully not have them crash around him. He went to school – he walked, didn't drive – and then got a job on a local newspaper. After that he moved to London where he worked at *What Car?* magazine, spending most of his time arranging fast and expensive cars for the publisher to drive at weekends. However, this hard work and dedication to duty paid off handsomely when three years later he was appointed Editor for the launch of *Which Car?*, a magazine he edited for six years.

Eventually he got bored with commuting, which on his journey consisted of sitting for two hours every day in slow-moving traffic. On one such static occasion, he worked out that by the time he retired he would have spent around a year and a half sitting unmoving on the A217, breathing in other people's car fumes. He resigned that very morning, went freelance and, amongst other things, started to compile a file of stories which would one day end up in this book.

He still spends most of his working day writing about cars but when he drives now, he makes sure he has his sunglasses on the right way up. This seems to have worked, as he has not had an accident. Yet . . .

MOTORING *SHORTS*

CONTENTS

Introduction 6

IN THE BEGINNING
The dawn of motoring 17

THE CAR MAKERS
They're mistake makers, too 35

ACCIDENTS WILL HAPPEN 40

ANIMALS IN CARS
Not a roaring success 50

THE PERILS OF PARKING
And the dreaded traffic wardens 53

THE LONGEST JOURNEY 70

AMAZING CLAIMS
Insurance pleas with a difference 72

COURT IN THE ACT
I put it to you 84

DON'T DRINK AND DRIVE
It's dangerous 94

FAMILY AFFAIRS
*Sometimes you don't know who
your friends are* 99

MOVING IN MYSTERIOUS WAYS 103

LEARNING TO DRIVE 106

POLICE SQUAD
*They're on patrol, but don't always
get their man* 111

TAKEN FOR A RIDE
Here come the car thieves 114

THE WRITE STUFF
A journalist tells tales 117

MOTORING  *SHORTS*

INTRODUCTION

Rolls-Royce almost called their new car Dung Heap, or at least that's what its proposed name – Silver Mist – meant in German. No, it's no joke. Even the mighty motor corporations make blunders, and sometimes they truly are of the giant kind.

But they're not the only ones. Motorists out on the road have their fair share of disasters too, and many get involved in some very dubious, not to mention dangerous, driving activities indeed. In this collection of Motoring Shorts you will discover drivers whose cars have floated away down rivers, drivers who have called God to give evidence on their behalf in a court case, drivers who have made a habit of running people over, a driving instructor who was involved in a high-speed chase with police, and a driver who had his car blown up by anti-terrorist police because he parked outside a General's house. They are all here.

I've divided the book into sections which deal with specific areas of motoring errors and events, but first of all, let's take a look at some of the more general boobs which have happened out on the roads around the world. Here's something to get you thinking . . .

The way you drive could be ruining your sex drive, according to English expert Dr Malcolm Carruthers (well he had to be English with a name like Carruthers, didn't he?). According to the doctor, 'Testosterone in men and oestrogen in women are reduced if you drive for too long, because of the stress, so it really can affect your sex life. Notice how you often don't feel like making love after a long drive. Now you know why.'

MOTORING SHORTS

During a test carried out by one of the UK's motoring magazines, a reporter was wired up to a heart and blood pressure meter and then asked to take a long journey. The results were alarming (in fact, enough to send your blood pressure soaring and your heart beating madly . . .). The driver's heart was beating at a healthy 60 beats per minute (55bpm is normal) and it maintained this until a car suddenly pulled out in front of him – then it shot up to 100bpm. Dangerous enough, but the 144bpm peak in this test came on the motorway with rain beating down outside, underlining the potential problems of driving long distances, especially when the weather or visibility is poor.

Professor Gary Cooper, a stress specialist at Manchester University, was not surprised by the results. 'The problem is that when you drive somewhere to a deadline you initially feel as if you are in control; but due to factors such as traffic hold-ups, you aren't, and the growing powerlessness is what causes so much stress.'

So now you know.

You would hardly believe some of the mistakes that British drivers make while driving in mainland Europe. Some are best excused as the result of forgetfulness, but others are just downright dangerous!

We know this because every year Britain's biggest breakdown organisation, the AA, keeps tally of the number and type of breakdowns its units attend from their centre in Lyon, France. There are the usual ones, such as engine overheating and clutch failure, but how about forgetting that traffic on the right has priority? Or filling up with diesel instead of petrol? Apparently, if you do this you can still drive on – but only for about 10 yards.

Another common piece of forgetfulness, which I suspect would quickly give your holiday a certain ambience, is someone overlooking the fact that Calor Gas is not available in France. This could make a caravan holiday on a tight budget something to remember. Then there are the hundreds of motorists who venture to the continent every year and lose their car keys, only to find they don't know the serial number so they can't get a replacement.

Another item AA Five Star continental insurance cover often had to pay out for was loss of tents, so it's perhaps worth checking you did pack it when you left that other French campsite last night – apparently hundreds do drive off without theirs. Perhaps they are the same ones who forgot the Calor Gas, and they decided to leave their tents as well so that their disastrous French holiday could be left behind and forgotten that much quicker.

Talking of being left behind, motorist Ian Harris waited so long for the RAC rescue service to come and get his car going again that he froze – ambulancemen found him sitting in his car suffering from hypothermia. Harris had rung the RAC five hours earlier when his car stopped on the M6 near Preston, Lancashire, in August 1989. He had to be rushed to a nearby hospital so he could thaw out. 'So much for the RAC saying they can get there in an hour,' complained Ian later. The RAC apologised and launched an inquiry.

Mind you, at least Ian Harris was eventually saved, unlike this chap. In 1969 an advert appeared in *The Times* which stated, '1928 Rolls-Royce Hearse, original body, excellent condition.' See, you can still get optional extras without always having to pay for them.

The ease with which Porsche-driving tennis star Hana Mandlikova passed a car being driven by an off-duty policeman at Eastbourne did not please the Sussex bobby one little bit. He raced to catch her at a set of traffic lights and wound down his window to demand, 'How come you passed me back there?' Of course, Mandlikova didn't know he was a policeman and thought he was just an obnoxious man, so she replied, 'Because my car had the power.'

The red-faced officer then said, 'I don't like you going so fast,' and Mandlikova replied, 'I can understand that: you don't have such a good car.' And off she drove, leaving the out-witted copper fuming. Next time, sir, if you ever catch her, perhaps it would be better to wear your uniform.

Here's a rather bizarre rendition of English, from a sign discovered on the wall of a hotel in Fontainebleau, France: CARS LEFT IN FRONT OF HOTEL ARE AT OWNER'S RISK. Okay, nothing much wrong with that. But it's the line underneath which is a bit worrying: GARAGE IN OUTER SPACE AVAILABLE ON REQUEST.

This piece of misunderstanding appeared in a Belfast newsletter in 1967. 'Bachelor (40), non-driver, would accompany same on car tour of Ireland.'

Hmm, presumably they've not set off yet?

One person who had set off, at least for a short distance, was Chinese motorist Chin Chi-Sing, for he had become the very first Ferrari owner in China. Problem, though. Unfortunately, the roads in China are so bad that Chin can only drive the super sports car around Beijing. Venture further afield and the car that can top 180mph comes to a grinding halt on the uneven road surfaces.

Traffic around Epsom in Surrey ground to a halt as simply too many cars tried to park near the world-famous racecourse. It took police hours to move all of the vehicles and to get traffic flowing again.

Sound familiar? Well, it could be today, but in fact it was a report on one of Britain's first ever traffic jams, which involved an estimated 45,000 cars and happened in 1928. Some things never change.

MOTORING   *SHORTS*

In June 1989 police forces across Britain carried out a swoop on foreign lorry drivers to see if they had had the right amount of sleep and stop-time. Officers were horrified at the state of some of the drivers they found. For example, one French lorry driver had had just three hours' sleep in 48 hours' driving, while Nottinghamshire police revealed that an Italian truck driver had rested for four hours out of 28! Superintendent Roger Storey said at the time, 'Some of them were just waiting to have an accident.' And probably by now, many of the stupid truckers have.

A rare Rolls-Royce, which had been sitting in a garage for so long it ran up a parking bill of over £18,000, was eventually sold in February 1990 for a massive £200,000 to an American car collector. The Phantom Six, coincidentally the Queen's favourite car, had been left at London's Sheraton Park Hotel, but it hadn't been abandoned by its previous owner – it was just hardly ever moved. The owner had even been paying the parking bill! The car was fitted with a TV and a fridge and was bought by the American through a London car dealer, who said that the US high-roller was an American-based tycoon who wanted to use the car when he visited London on occasional business trips. He already had several other Rolls-Royces in London for use when he visited. No wonder the roads are so crowded!

One of the world's most consistent offenders, American Ken Lock, celebrated his 100th speeding ticket in Richmond, Virginia, by giving away his car after deciding that his driving days were over. 'I have done this before I kill myself,' he declared.

Cabbie Pedro Zardoras, 74, offers passengers half-fare to drive themselves! 'My eyesight is not what it was,' admits Pedro of Cordoba, Argentina. 'I tell people they'll be safer at the wheel than me.'

National Westminster Bank blundered when they presented a funeral director with a bill for £1000 worth of bank charges, as he refused to take the matter lying down. He blockaded the bank's car park with his hearse, so no-one could get in or out.

His financial trouble had begun when he accidentally ran-up an unauthorised overdraft. He didn't know about it because the bank kept sending letters to his old address, even though he had told them he had moved. Eventually the charges incurred on a £600 overdraft had soared to £1000. The undertaker, of Marple, Cheshire, said, 'I could have got it cheaper from a loan shark. I just had to do something.' Nat West eventually agreed to refund £750 in bank charges.

I apologise for this in advance, but apparently women drivers are worse than men – it's official. Forgetting where they are going, missing motorway exits, failing to stop at crossings, ignoring Give Way signs, overtaking on the inside, and having more – but less spectacular – accidents than men . . . The facts came to light in a study carried out at Manchester University in August 1989. Dr Steve Stradling pointed out that 'women generally drive less than men and therefore get less practice', though he did add that a young man with his mates in the back 'is as lethal as a guided missile'. Well, that evens things up a little bit.

We all know drivers who change dramatically when they get behind the wheel. They turn from perfectly normal people into irrational beings who see the car – or in the case of this tale, the bus or van – as a weapon, something to be fought with and fought over. This particular episode from 1989 began when a bus and a van, coming towards each other, both tried to pass a parked car at the same time in the small Oxfordshire village of Eynsham.

The man in the van started arguing with the woman driving the number 90 bus, claiming that she should have given way as the car was parked on her side of the road. But she wouldn't have it, and after a heated row the van driver stormed off, leaving his locked vehicle jammed against the side of the 40-foot bus. By that time a further three buses had queued up behind the stranded single-decker, as well as assorted cars and vans. Seven passengers were caught aboard the number 90 – which normally takes 45 minutes to complete the journey from Oxford to Carterton but on this day took around three hours! – and one of them, out on a shopping spree with her mother, explained what had happened.

'It was obvious both drivers were doing it to make a point. They weren't interested in the passengers. The bus ought to have stopped because it was on the same side of the road as the parked car, but the driver said she couldn't reverse back as she would have been backing onto a main road. She was determined not to give way.' Instead of getting home at 1.30pm, the bus passengers eventually arrived shortly before four o'clock. 'Nobody apologised to us and I'll never ride on one of their buses again. I shall use the rival service,' said one, who added, 'I wouldn't have minded if they had said they were sorry; but it was quite disgusting, just leaving us sitting there.'

The Bus Company Operations Director, David Beaman, said, 'The driver is one of our most reliable. She says that she could do nothing after the van parked in front of her. The van driver apparently locked up his vehicle and walked off. It was only when our driver called the police that she was able to move.'

So, who finally had to give ground after the three-hour delay? Well, neither of them, as it happened. The police decided to move the parked car!

The Low Wood Hotel in the Lake District got so fed up with day-trippers stealing guests' parking spaces that in 1990 they advertised for a 'thoroughly nasty and vindictive' car-park attendant to chase them away. The hotel was deluged with replies for the £4-an-hour job; applicants included a man with friends called Basher

MOTORING  *SHORTS*

and Crusher, and a lorry driver who hated motorists. General manager John Gower said he was in search of someone who 'can stand up to abuse. We don't want wimps.'

I've not been up to see how they are getting on – personally I don't fancy the thought of being chased by Basher or Crusher – but if I do go to that area of the Lake District, I guess I'll go by train. Just to be on the safe side.

Prince Charles had a brush with death, or at least he came close to being an acquaintance, when he went for a spin in a Formula Two racing car at Thruxton circuit. The Prince was under the instruction of the late, great world champion Graham Hill – son Damon now carries on the racing family tradition in today's Formula One Championship – when it started to rain. As keen Grand Prix fans will know, a Formula One or Two car normally runs on slicks, the treadless tyres which grip exceedingly well in dry conditions but which allow the car to slide around like Nureyev if it rains. Graham Hill knew that he must try to stop the heir to the throne before he killed himself, so he leapt into the driver's seat of Prince Charles's own Aston Martin sports car and raced after him. 'I had no hope of catching him,' said Hill, 'and then I saw the car begin to spin. I thought, what have I done? This is the future King of England, I'm going to wind up in the Tower.' Fortunately, Charles didn't lose control of the 180mph car and Hill got to keep his head.

If you are something of a regular with the police – for speeding, that is – then you only have a couple of alternatives if you want to hang onto your licence. The first and best, I humbly suggest, is to slow down and abide by the speed limits, but another is to choose a less conspicuous car, because there's no doubt that the boys in blue like to stop the flashily quick. (I was once stopped three times in one day in a bright red BMW M3 for no sensible or proper reason; in one case it was just so that one of the officers could stare at the engine. Last

time I caught somebody looking quite so engrossed was when I saw American actress Michelle Pfeiffer crossing the street in London. Now I know some consider the BMW engine to be attractive, but really . . .)

Anyway, Viscount Linley, Princess Margaret's son, was no stranger to the police back in the late 1980s; he'd already been banned by the courts twice for speeding offences. So, in an effort to be as inconspicuous as possible, the Viscount swapped his flashy and very quick Aston Martin for a far more 'sedate' 1951 Morris Minor convertible. No danger of being caught speeding in this old dear of a granny's car: it might top 60mph, but only just . . . Wrong! This breathed-on version had been tuned by specialists, the suspension sorted out and improved, extra-wide wheels and tyres bolted on, so it was capable of well over 100mph. But what with all the press pictures and newspaper stories at the time, there can't have been many police officers who didn't know all about the new wolf-in-sheep's-clothing Linley machine.

Bad posture while driving is emerging as a major reason for the growing incidence of back pain, say osteopaths. A survey by The Guild of Osteopaths and other linked organisations found that eight out of 10 patients suffered back pain because of driving, with the problem three times worse for drivers at the wheel of a manually geared car, compared with an automatic.

Mum braking too heavily, Dad putting his hand on Mum's knee, and both of them arguing over which route to take: these are the driving habits that have children cringing in the back seat. Half of all teenagers questioned in a nationwide survey in early 1994 thought their parents' driving was very good – I guess some of the young ones might have been turfed out on the hard shoulder if they hadn't – and only 3 percent said their mother or father was a bad driver.

In a separate part of the survey where the adults were questioned, 35 percent of them admitted to hogging the

MOTORING  SHORTS

centre lane of motorways (I've met these people on the M25) and only 45 percent said they used the inside lane. Another interesting revelation from this survey by a leading insurance company was that no less than 60 percent of motorists admitted speeding on non-motorway roads over the last year, but 95 percent of them had got away with it, leaving just the unlucky five percent to face the music. Nearly half of all drivers, 47 percent, confessed to breaking the speed limit on motorways, with only 1 percent of them being stopped. Of 32 percent who admitted parking illegally, just 15 percent had been caught, and 20 percent of motorists questioned admitted to driving regularly without their seatbelts on.

Old codgers who are driving too slowly run the risk of being stopped by police officers and being told to speed up or get off the road! This campaign was launched in February 1994 on the Isle of Wight where oldies had been holding the traffic up – even during the off-season! – causing massive traffic jams which would literally end up blocking the island's roads.

A 16-year-old teenager in New York got bored with driving cars, so he hopped on a subway train and drove both it and its passengers 45 miles on the network. Keron Thomas fooled commuters by wearing a uniform and carrying an official-looking equipment bag. He took the controls of the train at Manhattan, drove on through Brooklyn and Queens, and had almost completed the return trip when he drove into a bend too fast, went past a red signal alerting him that he was speeding and triggered the emergency brake. In all he drove for two and a half hours on the longest line in the city, even making all his stops on time!

The young man was taken to Transit Police HQ for a routine alcohol test but escaped by running back into the subway. He was later arrested after police traced him through telephone records, which was only possible

MOTORING  *SHORTS*

because he had earlier phoned a crew office to volunteer for overtime, giving the name of a real driver but leaving his own home phone number.

Thomas was well known to rail workers because of his close interest in trains but rail staff and union officials were shocked at the ease with which he had managed to get behind the controls of a train. 'This is not like getting behind the wheel of the family Oldsmobile,' said Transit Authority spokesman Robert Slovak. 'You can't just walk in without any prior knowledge and drive the train.' Try telling that to Keron Thomas.

There's always been debate about the future of petrol- or diesel-powered cars and whether they will all eventually splutter to a halt as oil stocks diminish. But in fact, there's no immediate threat of oil disappearing, because each year since it was first discovered more oil has been found than has been used, and there are still areas of the world where the oil companies have barely scratched the surface.

Test drilling in the South Atlantic, for instance, has revealed that the area around the Falkland Islands contains at least as much oil as was discovered around the British Isles twenty years ago, and similar finds are being made in other areas. Couple this with the fact that although cars are becoming more popular all over the world, and parts of the world where cars were rarely seen are now home to increasing numbers of them, they are also becoming much more fuel-efficient, and it will not be long before most cars, even the very quickest, will be capable of an easy 100 miles per gallon, thereby saving on fuel again.

Rather, the biggest threat to the oil-fired internal combustion engine is coming from the environmentalist lobby, but there is still a reluctance to hand over the keys of the petrol-engined car in favour of the cleaner alternative technologies which, as we'll see when talking about the Doble steam-car in the next chapter, have been with us for some time now.

MOTORING  *SHORTS*

IN THE BEGINNING
The dawn of motoring

The early days of motoring saw some amazing contraptions, a host of dangerous drivers including some complete buffoons, as well as some spectacular accidents. Here are some of the most amusing tales from those pioneering days.

In the early days of motoring, frightening horses with your car incurred huge fines. The first American to fall prey to this was Philip Hagel, who was the first motorist to drive up New York's Broadway in his French de Dion. He startled so many horses that by the time he'd reached the end of Broadway he faced damages totalling $48,000, a hefty sum today let alone back in 1906.

In 1895 the motor car still had no generic name, so a Chicago newspaper organised a $500 prize competition to find one. The winning name? Motocycle. Oh dear.

It wasn't until four years later that the French came up with the word automobile which the New York Times described at the time as being 'half Greek and half Latin, and so near to indecent that we print it with hesitation'.

One of the earliest blundering drivers was the wealthy American William Kissam Vanderbilt, who had the unhappy distinction of killing at least one living creature every time he got behind the wheel. So not surprisingly, on a journey across Europe between 1899 and 1908 Vanderbilt caused havoc. His journal reveals some of the incidents.

'On arriving in Frejus, killed two dogs that viciously attacked our tyres.' Unfortunately this was not the most serious episode. Just outside Florence a six-year-old boy ran in front of the Vanderbilt car and the American ran him over. Vanderbilt was set upon by an angry crowd

MOTORING  SHORTS

who only retreated when he took out his revolver and threatened to shoot them. He hid in a closet in a shop and was later rescued by the police.

Needless to say, Vanderbilt was not popular in Europe, mostly because it was his avowed policy never to stop after any 'incident'.

In 1897, arrogant Henry Sturmey, an early British driver, panicked a small donkey-drawn cart when he drove past at high speed, causing the woman driving it to lose control. The cart careered down the road with the driver fighting frantically for control. 'I confess I could not resist the temptation of an occasional blast upon the car's trumpet to keep the fun going,' chortled Sturmey.

But on this occasion the hooting Henry had made a bit of an ass of himself, for the donkey broke free of its reins, turned around and, braying loudly, galloped at Sturmey's car, causing the Englishman to panic and drive off the road and into a stream, where his car became wedged tight.

Some early cars were design blunders in a class of their own. One of the best examples of this was the elaborately named Kissing Bug Turnout, which was owned and driven by wealthy Frank Belmont, one of the United States' 10 richest men. In October 1899 Belmont's car was disqualified from the Automobile Festival at Rhode Island for one very simple but important reason – it was found that it could not be steered!

One of the world's worst learner-drivers was American Mrs Stuyvesant Fish. *The Automobile Magazine* reported in 1899 that, 'though there may well have been some portions of the well-kept grounds surrounding her house that were not ploughed up by the wild rearing and charging of her automobile, careful scrutiny failed to locate them.'

MOTORING  *SHORTS*

When the German engine-inventor Gottlieb Daimler was developing one of the first internal combustion engines, he very nearly ended up in court on forgery charges. As Daimler and his assistants worked on the engine in 1885, they were suddenly raided one night by the local police who mistakenly thought that the late-night cacophony of machines meant that Daimler was forging coins!

Fortunately the mistake was soon recognised and the great man was allowed to get on with inventing the motor car, ultimately an altogether more lucrative business than forging coins, though at times it must have seemed like a licence to print money as Europeans flocked to buy the new 'horseless carriages'.

The development of the earliest petrol-engines was carried out in small workshops and garden sheds in various European countries in the late 1880s, but the budding engineers had to keep their revolutionary work very secret indeed. Not because they thought their inventions would be stolen, but because local people often objected to such engineering feats being carried out: they feared the threat of fire brought about by these mysterious machines.

Many road races were organised in Europe and America as the car began to gain popularity in the early 1900s. One of the first casualties of such a race was Louis Renault's brother Marcel who would not wait for the dust to clear before racing into a sharp bend just south of Poitiers in France. Unable to see, the luckless Marcel careered off the road and was killed.

Louis, founder of the Renault car company, made a major misjudgment himself when he befriended the invading Germans during World War II. At the end of the war, he was jailed for collaboration and his company taken over lock, stock and barrel by the French Government.

MOTORING *SHORTS*

There were some intrepid explorers in those early days for sure, but not all of them came back to report on what they had found. It's hard to say exactly where American John D Davies and wife went wrong on their intrepid coast-to-coast drive from New York to San Francisco back in 1899, but go wrong they most certainly did. Why? Well, they left Detroit in mid-August . . . but neither they nor their car were ever seen again. Admittedly, cars were rather slow back then, but surely not that slow!

One of the first attempts to drive a self-propelled vehicle was not without its dramas. Way back in 1770 in Paris, France, Nicholas Cugnot drove his wood and metal fardier – designed primarily to tow gun-carriages – into a small square in one of the well-to-do suburbs. He drove the fardier via a large tiller-like device such as you would normally see on a ship. So, the vehicle came into the square, much to the alarm of bystanders, it crossed the square, and it went out of the square the other side, demolishing someone's brick garden-wall on the way! Imagine the faces at the insurance office the following week when confronted with <u>that</u> story.

When the first long-distance motoring journey was made by a vehicle driven by a petrol engine, it was not without its mishaps. The intrepid pioneers were Karl Benz's wife Bertha and the couple's two sons, who secretly took off on history's first joyride (no handbrake turns, this was the original 'joyful ride' – but they did not tell Mr Benz, they just went out for the day in 1888 to see how well, or otherwise, the world's first proper car went).

It was an eventful journey to the nearby town of Pforzheim where Frau Benz's parents lived. During the 60-mile journey they had to find water to replenish the cooling system, visit an apothecary for petrol, and push the car up one series of hills because it didn't have the power to carry all three of them up. They also had to stop to persuade a cobbler to make them a new leather brake-block and, one of the more personal events, they had to use one of Frau Benz's garters to replace a broken rubber insulator.

When Henry Ford, the world's very first billionaire, eventually got his car-making company properly into operation (previous Ford ventures had ended in bankruptcy and it wasn't until the great man was in his 40s that business began to boom), it was producing so many cars that one newspaper reporter of the day remarked, 'Two flies can breed 48,876 new flies in six months – but they haven't got anything on two Ford plants.'

In France, the relatively good roads, wide-open spaces and small population (I mean there weren't many of them) meant motor races soon became very popular indeed. The four-wheeled pioneers could race along for the most part to their hearts' content. However, there were serious mishaps, not least the regular deaths of drivers, as we have seen.

Just such a race was the Paris–Madrid of 1903 which was stopped by order of the French Prime Minister, Emile Combes, who banned the drivers from going any further than Bordeaux. Not only was this annoying for the competitors – those still alive, that is – but they had to suffer the embarrassment of having their cars towed by horses to the nearest rail station, from where they were transported on wagons back to Paris.

When he wasn't busy being general-secretary of the Communist Party, Soviet premier Leonid Brezhnev was a real car nut. He owned a small collection of 13 fast and expensive cars, including two Rolls-Royces, a Mercedes-Benz, a Cadillac and a Citroen-Maserati. Sadly, his driving ability did not match his enthusiasm.

'When I am at the wheel, I have the impression that nothing can happen,' announced Brezhnev grandly during a state visit to West Germany in 1973. Minutes

later he took the wheel of a Mercedes-Benz . . . and promptly crashed into the nearest tree.

If you were a dog, it was well worth steering clear of any race organised on French roads during the early 1900s, whether or not Mr Vanderbilt was around. Many drivers purposely took off their exhaust silencers so they could make as much noise as possible, and unfortunately this drove French dogs barking mad. So much so that in one national race run in 1900, the winner, one Henri Charron, killed no less than five dogs, all of which threw themselves under his wheels, apparently maddened by the raucous sound of his car.

But the dogs got their revenge the following year when Charron again took part in the race. Just south of Lille, a large mastiff ran out and viciously bit Charron's left front-tyre, causing a puncture which in turn caused the Frenchman to spin off the road and crash into a deep ditch. It didn't end there either. The dog then ran across the road, down into the ditch, and bit the unhappy Charron, causing an infection which kept him in bed for weeks. For the dogs of France, honour was satisfied.

Many automobile engineers and designers have tried to look into the future, designing cars which are far ahead of their time. Just one such design was the Dymaxion, unveiled in 1927. It was an 11-seat three-wheeler which could cruise at 120mph. Unfortunately the Dymaxion's inventor, R Buckminster Fuller, made a blunder of the most elementary kind. While out driving one of the few production Dymaxions in Chicago in 1933, the inventor foolishly got into a race with another car. The Dymaxion crashed and was written-off. After that, the project died.

Interestingly, the steering gear for the ill-fated Titanic was designed by Andre Citroen of French car fame. Though after the great liner had gone down, Citroen drily retorted, 'But the iceberg was American.'

MOTORING  *SHORTS*

It was two o'clock in the morning, and a light rain had begun to fall over Detroit. Inside the small brick shed four very tired young men were making final adjustments to the car's engine and fuel supply. They'd all worked virtually non-stop for the past 48 hours, but it had been worth it, for at last all was ready. Now the car's designer and chief builder climbed aboard. He made himself comfortable on the small seat, smiled down at his colleagues, and pushed the starter button. The little two-cylinder hand-made engine duly spluttered into life.

But with mounting horror, they all realised there was a problem, and it was such an obvious one. The car had been built from scratch inside the brick shed and they'd all been so intent on getting it just right that they'd forgotten about the shed door, which wasn't wide enough to drive the completed car through!

The driver knew what to do. He stepped down, picked up an axe, demolished the door frame, knocked out bricks and masonry until the gap was just wide enough, then got back aboard and drove the Quadricycle out into the night. The driver's name was Henry Ford and this was his very first car.

Just goes to show that even the world's greatest don't get everything right the first time.

One of the more bizarre motor vehicles of the early years was the spring-powered machine which supposedly worked by storing up power in a massive spring when the vehicle hurtled downhill, then releasing it as the vehicle moved along flat ground or uphill.

(Presumably this wouldn't have been much use in Holland, unless you could build a massive ramp outside your house and have the car rocket down that prior to each journey.)

Of course, such a device did not take off – unlike many of the early cars fitted with such a system, which quite literally became airborne, along with their drivers. Just imagine, you would have needed a very efficient braking system to stop the forward momentum of the spring, because once unleashed, the energy was terrific.

MOTORING *SHORTS*

Some of the earliest car races were hardly the frantic affairs that modern Grand Prix are, but pedestrians casually strolling in amongst the racers? Surely not . . . Yet this is exactly what happened when one of the first long-distance reliability trials was conducted in France in 1894. The idea was to drive from Paris to Rouen and it certainly proved a popular draw for spectators and competitors alike. Tens of thousands turned up to line the route, and there were 102 contestants, though in the event only 21 got their vehicles to the start line. Thirteen of them were powered by petrol, the rest by steam. During the 'race' itself, where heady speeds of up to 10mph were reached, bystanders were able to stroll in and out of the traffic and even walk alongside the drivers and chat to them as they raced for the finish line. All of the petrol-fuelled cars got to Rouen, while the others, well . . . they just seemed to run out of steam.

Some of the early motoring pioneers were truly heroic – or stupid, depending on your viewpoint. A case in point was Frenchman Emile Levassor, who was known far and wide for his early racing exploits. In 1895 he took part in one of the first long-distance high-speed races from Paris to Bordeaux and back, a fearsome 740 miles – tough even by today's standards. Levassor took only 48 hours to travel the route, and this was in a Panhard which averaged 15 miles an hour!

So how did he do it? Well, by driving twice round the clock. (We have already seen a French lorry driver doing much the same, ninety years on. In Levassor's day, of course, it was not illegal, even if it was downright dangerous, and Levassor was not driving a juggernaut.) He took only one short nap, as well as occasionally stopping for petrol and water, and he also consumed a glass of champagne before turning his car around at Bordeaux and heading back.

The earliest vehicles most often rode on what were basically cart wheels, sometimes strengthened to take the

MOTORING  *SHORTS*

extra weight, sometimes simply unbolted from a cart and bolted onto the car. Even those cars with custom-made wheels rode on solid tyres.

Help was at hand, however, from the Michelin brothers who helped develop the pneumatic tyre. In France this new-fangled device was fitted first to a Peugeot which gained the nickname Eclair (not a sticky bun with cream in the middle, at least not in this context – the word means lightning). Apparently the name was not earned because the car was fast: it certainly was not, a lowly 15mph being the top speed. Rather, it was because of its new tyres which caused the vehicle to zig-zag up the road – just like a streak of jagged forked lightning. Geddit?

If you watch a Grand Prix race today, you can marvel at how the cars have been adapted to be more aerodynamic and therefore quicker than their rivals – special spoilers, air ducts which flow the air over the body, rear wings which disperse the air cleanly. But back in the early 1900s, making a car more aerodynamic to win a race didn't occur to the competitors. There were considered to be just two ways of making the car go quicker.

The first was increasing the size and power of the engine, and the second, in tandem with the big engine, was having as low a weight as possible. But precisely because the engines were getting steadily bigger and more powerful throughout the early 1900s, the race organisers began to think of ways of making the machines more comparable to ensure a level playing field. There was another consideration: that of safety. Because the engines were getting bigger and the cars could go more quickly, they were, ultimately, becoming more and more dangerous.

For the gruelling Paris-Vienna race of 1902, the race authorities decided there should be a weight limit of 2200lbs (1000 kilos), which, they reasoned, would mean the bigger, much heavier engines could not be used. Wrong!

What the racers did was drill holes into just about

every bit of the car's structure so that the weight dropped, jettison just about everything which wasn't screwed down or which wasn't an integral part of the car's body, but keep the large engines. The result of this was that speed was up, the cars' stability suffered, and accidents soared.

One man who did make it to the finish of that 1902 Paris-Vienna race was Marcel Renault, but he arrived there so early that he had to persuade race officials that he was indeed M. Renault, competitor in the race. He arrived over half an hour ahead of the rest of the field, driving his 4.5-litre handmade car to victory.

There were some bizarre bits of advice, some of it enforced by the police and the courts, about how to drive those early cars. In Britain there was much concern about the effect of the noisy car on the large horse population, so much so that one horse writer gave the following advice to car owners: 'Drive the vehicle around the horse in ever decreasing circles until the beast gets used to the noise.' Presumably by beast he meant the horse, but at that time there were many who considered the car to be the beast.

When the car first made its appearance in the USA, home of the cowboy and his horse, many rural towns and legislators attempted to keep car owners firmly in the minority. One of the ways to make sure that the horse did not take a back seat, as it were, out on the road, was to give it precedence over the car. Therefore in 1901, the following requirement was made of car drivers in Kentucky: 'When an automobile encounters a horse in the countryside, the automobile must be completely dismantled and the parts placed hidden in the nearest hedge, until the horse and rider have passed; only then should the automobile be reassembled.' So maybe if you look closely enough you may still find bits of the early

MOTORING   SHORTS

Fords around Kentucky – if you can still find any of the hedges, that is.

In some states drivers were told they must fire rockets into the air as they approached a town, thereby giving residents, and horses, notice of their arrival. Personally, I'd have thought the rockets would have been more of a shock to the residents, both horses and people, than the arrival of a car, but there you are . . .

In Britain in 1903 the speed limit was raised to a meteoric 20mph and though this was welcomed by the motorists, of which there were an ever-increasing number, it was not welcomed by pedestrians, horse-riders, or those who lived by the side of roads and had to suffer the increasingly large clouds of dust kicked up by these new cars.

The police were brought in. Stop and prosecute the speeders, they were told. They carried out their new role with the utmost energy, stopping motorists and booking them with monotonous regularity. One of the speed-traps, if you can call it that, was between Brighton and London, Brighton then being a very fashionable resort regularly visited by the Royal Family among others.

The motorists soon tired of the attentions of the local constabulary and so started to organise their own police-spotting squads. These were usually groups of hired cyclists who would go down the route first, looking for the hidden police officers, and then report back to the drivers, who would then, given a clear run, speed down to Brighton, or, if there was a speed trap, presumably head further up the coast, perhaps to Worthing instead.

All of this palaver does of course beg the question, why not go by bike in the first place? Well, back then the car was still exciting, cycling was pretty boring and the car ultimately got you where you wanted to go quicker than a pedal bike would – though it wasn't much quicker if the car driver stuck to the speed limits!

Funny how things change; now, thanks to today's traffic congestion, it is often quicker to go by bike!

MOTORING SHORTS

One of the main reasons for the setting-up of the Automobile Association (AA) was in fact to warn members of speed-traps up ahead. The AA cyclist – no, they didn't have yellow vans then – would cycle along and if he spotted a police speed-trap he'd cycle back up the road and warn all those approaching motorists who sported the AA badge on their car's bonnet.

Trouble was, this eventually was construed as obstructing the law, so the AA changed their stance slightly and issued a guideline to their members which said, 'If your AA patrolman does not salute an approaching member, stop and ask him the reason why.' Clever. Presumably the AA man would say, 'Sorry I didn't salute you there, got a stiff arm, played bowls for too long yesterday. By the way, some chaps in the bushes up the road there, blue uniforms, funny hats. Now good-day to you, sir, and mind how you go.'

Rolls-Royce have always had a reputation for smooth running and refinement, a feature which was there right at the start of production in 1905. So quiet was the engine of the famed Silver Ghost that the *New York Herald* newspaper reported that this was one of the first cars fitted with what they called 'a glass tell-tale' (a dial to you and me) to tell the driver if the engine was running or not. This might be something of an exaggeration, because even the Rolls-Royce was not that quiet (ask any horse . . .), but it did point the way ahead. Back then, as much as today, refinement as well as sheer speed would very often be the key to sustained sales.

You may think it's just American kids and rock stars who wear their caps backwards, and you could certainly be forgiven for thinking it's a modern look. But it was those early motorists who pioneered this street-cred style: they discovered that they had to put their caps on backwards, otherwise they got blown off as the wind rushed into the open cockpits.

I have here a description of a Parisian taxi driver. 'There is no escape if you enter his car. He lights his cigarette, sinks back into his seat, and his shoulders into his back, and his head into his shoulders, and drives like the devil. He seems to have no life of his own at all, he merely exists to urge the car forward wherever he is told. The foreigner has no hold whatever, the driver arranges the meter to whatever tariff he pleases, and before you can examine the dial at the end of the journey, he has jerked up the flag.' This ample description of the charms of travelling by taxi in Paris was written in 1912 by one EV Lucas, though I'm sure I met this same taxi driver only last year . . .

One of the major steps forward in automobile driving can be put down to a man who designed cash registers. American Charles Kettering invented the self-starter, which basically worked on the same principle as the cash-register, giving the engine a short burst of power of the kind used to shoot the cash-drawer out. No-more of that manual engine-cranking, which not only required a particular knack but was often quite dangerous, because the engine could whip the heavy starter-handle back all of a sudden, with enough power to break a man's arm.

The world-famous Rolls-Royce bonnet emblem – the Spirit of Ecstasy – was only designed and placed on the bonnets of the Rolls-Royce because the men at Rolls were increasingly distressed by the bizarre, sometimes obscene, mascots and emblems that owners often stuck on their cars. These ranged from massive gilded eagles, wings fully outstretched, to donkeys, parrots, fat policemen, copulating pigs (just the two of them though, apparently the idea of a pig-farmer from Dorset), and copies of family dogs and cats.

Thankfully, at least none of these appeared to be the stuffed remains of favourite pets, though an American farmer from Fayetteville, Arkansas, one Elmore Passmore, did come close. He wanted to put a stuffed

MOTORING  *SHORTS*

version of his late-lamented cat Oscar on the bonnet of his 1902 Curved Dash Runabout, built by Ransom Olds, father of the Oldsmobile company. The Fayetteville town council got to hear about this and banned Passmore from the act, because they feared it could set a dangerous precedent!

In October 1950, a 'slow' race for motor cars was organised along Rue Lepic in Paris. The winner, M. Durand, took 10 hours, 40 minutes and 51 seconds to drive a total distance of only 722 yards. Timekeepers all along the route carefully watched the snail-like competitors, ready to disqualify any motorist who actually went so slow that he stopped. They noted that at one stage Durand's wheel took nearly three minutes to make one complete turn.

Two years later, another motorist was slower still. He took 12 hours to travel only 49.5 feet. Regular M25 users will know the feeling.

There were some really bizarre attempts at producing high-speed cars, but truly one of the best, at least from a curiosity point of view, was the rocket-propulsion car which the German Opel company secretly tested in 1928. This machine had no less than 24 rockets in its tail and they powered the car by detonating at intervals as the machine gathered speed. The best speed it ever reached on a racetrack was just over 125mph, but of course once the rockets were burnt out it couldn't go any further.

There's plenty of talk about car aerodynamics these days, and a car which cannot slip through the air with ease – while keeping its wheels firmly on the ground, of course – is a car which is unlikely to sell. In the United States in the 1930s, cars looked great, but they were about as aerodynamic as a house-brick. A study at the time, carried out by airflow experts, discovered that most American cars were more aerodynamic going

MOTORING  SHORTS

backwards than forwards! This was useful, though, because it did start the debate about smooth and uncluttered airflow, which today helps give good fuel economy as well as good performance.

In the modern age of concern about the environment there is much research into alternative energy sources to power today's car. But in the 1930s the Americans already had the answer – the modern steam-car.

Now this is not as silly as it sounds. A company called Doble – named after founder Abner Doble – did produce a car which worked on steam, powered effectively just by water. In the morning you pushed a button and the starter lit a burner. This then took about a minute to warm the boiler sufficiently and then the car could be driven out onto the streets. It was virtually silent – though apparently it smelt like a Chinese laundry on overtime – and it could cover 1500 miles on 24 gallons of water. Power was from a straightforward four-cylinder engine which produced 75 bhp.

The problem for Abner was that he could not afford to set up a mass-production facility like his contemporaries Ford and General Motors, and so the Doble was very expensive (though interestingly it was guaranteed for three years, the sort of warranty which only a few car-makers have the confidence to offer today). Eventually, the company failed.

So why doesn't somebody make a steam-car today? Answers, please . . .

In Britain in the 1930s, driving on the roads was a dangerous business, not least because the 20mph speed limit was abolished early in that decade. This led to some hair-raising driving and led one motoring commentator at the time to say, 'The safest way by far to deal with a crossroads is to go flat-out, so one spends as little time in the danger area as possible.'

MOTORING   *SHORTS*

The company which became world famous as Jaguar started life as Swallow Coachbuilding in 1933, but was later changed to SS Cars. In turn, the name SS Cars had to be changed come the start of World War II for obvious reasons. Jaguar, it seemed, was less threatening.

While on the subject of World War II, one of the most unreliable pieces of machinery was the German Panther tank, circa 1944. When it decided to work it could blast just about anything it wanted into very small pieces, but the trouble was, more often than not it would just stop still in its tracks. So, German engineers didn't always get it right!

When the new Citroen 2CV first saw the light of day in 1949, a leading French newspaper of the time called it 'an abortion on wheels'. The car had been developed from an outline by André Citroen who had a set of demands for his new car. It had to have a tall enough interior so that a man wearing a top-hat could drive it, it had to have soft enough suspension to enable it to be driven over a ploughed French field while carrying six dozen eggs, which had to reach the other side of the field without breaking. Well, we all want something different from our cars, something individual, but . . .

There were some bizarre early contraptions in the motoring world for sure, not least the gas-powered Morgan car which had a short life in 1918. It was launched because of a petrol shortage in Britain during the First World War, and consisted of a Morgan car (of course) but with a big frame like a four-poster bed on top, and placed inside this cradle was a large bladder-like bag which was filled with gas. Today, a device of such design would be enough to get the fire-brigade rushing to the scene – what an extraordinary fire hazard it must have been!

But the system would probably have lasted for quite a

MOTORING  *SHORTS*

while, had not the war ended. It was designed and made by two men called Mr Edwards and Mr Perry (history does not record their first names) who owned a garage in London's Great Portland Street. The frame they constructed for the Morgan supported a gas bag measuring 7ft 10in long, 4ft wide, and 3ft high. Can you imagine?! This bag held enough gas for 30 miles of driving and the contents cost about a shilling (5p to you decimal children). It took about 10 minutes to fill it with gas. A half-inch flexible gas pipe connected the bag to a carburettor and the flow of gas was regulated first by a handcock, and second by a butterfly valve connected to the accelerator pedal.

Thankfully the German generals threw in the towel and the war ended, so we were able to get petrol again and the Morgan gas car was no more.

Pride cometh before a Hall fall: motorist JJ Hall, one of the pioneers of motor car racing, published a letter in 1927 expounding the virtues of the three-wheeler Morgan, saying that he had covered over 1600 miles at the Brooklands Racing Circuit in Surrey – then one of the world's major race circuits – without any problems and 'without ever having the slightest anxiety about crashing'.

Yes, you've guessed it – only a couple of weeks later, JJ Hall had a serious accident at Brooklands following a rear-wheel puncture. And as these early Morgans only had one wheel at the back – two at the front – this was something of a problem!

This last story is not so much In the Beginning as A Long Time Ago. Mechanic David Wall is still hunting a driver who left a classic Mercedes-Benz convertible at his garage – nine years ago! The 1960 sports car would be worth around £30,000 if it was restored, but instead it has just been gathering dust at David's garage in Norwich. Unfortunately, the car has no number plates to enable its owner to be easily traced. Now the mechanic has called for the owner to come forward – before he sells the car. 'I can remember what he looks like,' said David, speaking in March 1994, 'just in case anyone tries to pull a fast one.'

MOTORING  *SHORTS*

THE CAR MAKERS
They're mistake makers, too

You might think that when a car maker designs and eventually manufactures a new car, all would be well. And why not? After all, developing a new car from the drawing board, to the showroom, to your front driveway, costs a fortune; so it makes sense to get it right. As an example, just a slight redesign of an existing car can cost in the region of £360 million, while a completely new car like the Ford Mondeo, the company's most ambitious new car since the Ford Model T, has cost around a billion pounds.

Yet despite all the money these companies spend around the world, sometimes they get it wrong, and sometimes it's in rather a big way. The most common error is to give a new car a name which means one thing in one market, but means something altogether different in another . . .

Toyota's mid-engined sports car, the MR2, has been a major success in just about all markets with the exception of France, where only a handful have been sold. Toyota's Japanese management could not understand the sales flop until someone told them that MR2 in French is pronounced 'merde', which is French for the word 'shit'!

They still sold 85 MR2s in France in 1993. Can you imagine?

'Bonjour, Pierre, what is that sleek, smooth, sports car you are driving around in?'

'What? Oh yes, it's a Toyota.'

'Which one?'

'Oh well, you know, it's the Toyota MR2.'

'Merde!'

'Yes. But it is very quick.'

MOTORING  SHORTS

More problems for the Japanese. When the Colt Car Company, UK importers of Mitsubishi cars, received their first shipment of turbocharged sports cars from Japan, there was still no European name for the car. As is the way with these things, time was limited, the cars were here and Colt wanted to start selling them.

Rather than make a name up themselves, they contacted Mitsubishi head office on the phone and after some deliberation Colt were told what the name of the new car should be, so the badges were made and stuck on the car.

When the Japanese visited England some time later they could not understand the name, Starion. Slowly, it dawned on the British what had happened. The Japanese had indicated over the phone that they wanted this high performance Colt to be called Stallion, but unfortunately no-one in London had made allowances for the Japanese pronunciation of Stallion as Starion!

One of the most unsuccessful cars in recent times was the Fiat Argenta, a large four-door family saloon. It was well-priced, well-equipped, practical and reasonably attractive. So what could have been the problem?

Well, it cannot have helped that the Argenta was launched in Britain at around the same time as Argentina invaded the Falkland Islands. Unfortunate timing, Fiat.

Even the mighty can fall. America's General Motors, the world's biggest company, should, in theory at least, be home to some of the world's top marketing brains. How was it, then, that when they launched their European small hatchback car in the 1980s they called it the Nova?

When the car was launched in Spain, traditionally one of GM's most lucrative markets, sales of the new car were disappointing. Could it have had something to do with the fact that in Spanish 'No va' means No go? Oh yes.

MOTORING  *SHORTS*

In the period after World War I, car makers across Europe got ready for what they hoped would be a big sales boom, and some of the more enterprising companies decided to try their hand at export sales – until then virtually unheard of.

One of the attempts which did not go quite as planned was Fiat's big sales push on the new Super-Fiat, a V12-engined monster. They sold five.

It's red, it's low-slung, it sounds delicious and it goes as quick as a missile. Of course, it's a Ferrari. For real sports car enthusiasts there is nothing like it – the prancing horse emblem on the bonnet side, the growling exhaust, this is <u>the</u> sports car.

However, Ferrari made a bit of a blunder when they designed their cars – there's no room for a spare tyre! If you are unlucky enough to have a puncture in a Ferrari there is only a thin 'space-saver' tyre to replace the damaged tyre and wheel with, and more seriously, there's nowhere to put the cumbersome, not to mention expensive, damaged tyre and wheel. Actually, this is not strictly true: it can be placed on the passenger seat.

Unless you have a passenger. And as every red-blooded male knows full well, a Ferrari without a young lady aboard is not really a Ferrari at all. Oh dear. Could this have been the end of many a beautiful relationship?

'Didn't you bring a spare?' she says angrily.

'Well, there's no space for one,' he says sheepishly.

'I've never known such a design blunder,' she says. 'I'm walking.'

Bang goes the relationship, like the tyre – but at least you can get the damaged tyre back on board.

The suspension on a Morgan has always been hard. So hard, in fact, that in one notable road test in the 1960s when Morgan slightly changed the suspension arrangement, the writer said it was now possible to tell if

MOTORING  *SHORTS*

the cigarette end you had driven over at high speed had a filter or not!

Somewhat earlier than this telling remark, consider the sheer resilience of one Gwenda Stewart who made a real name for herself driving Morgans in various races and hill climbs in the late 1920s. On several occasions, such was the harshness and unforgiving nature of the Morgan suspension that when she was taking part in timed hill-climbs on less-than-smooth surfaces, holes would be worn in her overalls. Ouch!

It may seem strange now, but the first production car made in Japan was driven by steam. Built in 1904 by one Torao Yamaba, it had a two-cylinder engine, developing a minuscule 25bhp. On the face of it, steam was a strange choice of power because by then petrol engines had been around for 15 years or so and had started to sell well: in France, then the world's leading car-making nation, some 60,000 cars were already sold and out on the roads.

But Yamaba didn't intend his first car to be a mass-produced vehicle. It was simply meant to cart his substantial family around Tokyo, so it had 10 seats, was 14½ feet long and 4½ feet wide.

The Japanese may produce some worthy cars, and on occasion they produce some which become very well thought-of indeed, but as we have seen the same cannot always be said of the model names they choose. Nissan have produced arguably the most embarrassing, or at least the most insipid of car names: Sunny, Bluebird, Cherry . . . yuk! And how about the Cedric, redeemed only by being not quite as bad as the Mazda Bongo.

Other names which as far as anyone can tell mean nothing, yet are often hard to get to grips or identify with, have included: Daihatsu Fellow Max, Daihatsu Taft, Daihatsu Atrai; Honda Life, Honda Quint, Honda Today (so far no Honda Tomorrow or Honda Yesterday).

Then there's the Isuzu Bellett, Isuzu Rodeo Bighorn (also sold in Australia as the Isuzu Kangaroo, which to

MOTORING  *SHORTS*

me, at least, suggests only that it jumps all over the place), Isuzu Aska; Mazda Carol, Mazda Cosmo, Mazda Capella, Mazda Chantez; Mitsubishi Delica; Nissan Liberta Villa; Subaru Alcyone; Toyota Sprinter Trueno.

A particular favourite of motoring journalists was the Toyota Crown, a great barge of a car stuffed with just about every automotive gadget known to man – it even had a cooler compartment for drinks and sandwiches – but universally known as the Toyota Clown, ever since Japanese executives were heard talking about it in English and pronouncing Crown as Clown!

The old Soviet Union was not renowned for its car design and production, so it was perhaps fitting that the first export of Soviet-built Moskvitch cars to Norway was exchanged for a boatload of fresh herrings.

Personally, having driven a couple of Moskvitches, I think the Russians got the better part of the deal.

When the Datson car company was formed the name soon had to be changed. While the word Dat means hare in Japanese, and this was reckoned to be a good association for a fast-moving car company, when Dat was put together with Son (it was meant to mean literally 'son of Dat', Dat being the original company) the word sounded suspiciously close to the Japanese for Ruin!

So, they changed the name slightly to Datsun, the difference in sound (<u>What!?</u>) apparently giving an entirely different meaning in Japanese and being seen as an invocation to the Sun for protection.

The things people say, number 94: 'Safety does not sell.' The speaker was Henry Ford II, speaking over 20 years ago. Today, all car-makers recognise that safety and security are in fact top of the agenda when people are buying a new car.

MOTORING  *SHORTS*

ACCIDENTS WILL HAPPEN

Indeed they will, and though 99 percent of accidents are caused by people, car drivers don't always like to blame themselves . . .

The first recorded death of an American motorist was just plain careless. The man killed was one Hieronymus Mueller and a contemporary report reveals what happened.

'While filling the gasoline tank of his automobile, he came too close to a match which ignited the gasoline and set his clothes on fire.' Well it would, wouldn't it?

Veteran motorist Herbert Brooks drove into a garage for a £10 repair and left the building in a state of serious disrepair. A former steam-roller driver, aged 88 when the incident took place in 1990, Herbert roared onto the forecourt at more than 40 mph and managed to completely wreck two cars, seriously damage two others and write-off a fridge freezer.

To be exact, he crashed into a truck, bounced off into a Mini, shunted it into a Toyota Land Cruiser, then smashed into the workshop, demolishing the fridge. 'I think I got my gears stuck,' Herbert told staff at the garage in Market Harborough, Leicestershire, and added: 'I won't be driving again, it's far too dangerous.'

And not just for Herbert, it seems.

Blackpool Airport's a pretty quiet place, at least by the standards of the main international terminals. In fact nothing much ever happens at Blackpool Airport. That could well have been the thought in air traffic controller Clive Mackrell's mind as he gazed out of the control tower window in May 1993 and watched a lone four-seater Cessna coming in to make a landing.

He watched it come down lower and lower. He watched it suddenly dip down and touch the runway, then try to lift off again, then veer sideways, then . . . hit the offside of his Rover 213 parked near the control tower! After slicing its propeller through his Rover's bootlid, and caving in part of the car's roof, the Cessna eventually came to a halt.

'A deathly silence followed before it sunk in that the plane had actually crashed into my car,' said a stunned Clive. He'd been planning to sell the car the following week, but after its collision with the plane it was a write-off. 'It's not every day you get to write "hit by plane" on your insurance form,' said Clive.

Motorist Andy Puddifoot didn't pay a lot of attention to the sign which warned of sudden high tides as he parked his gleaming new Alfa Romeo by the riverside at Twickenham. But when he came back an hour later, the river had risen so high that his car had been lifted up and swept downstream, passing boats, two bridges and several fishermen before becoming firmly wedged in the locks at Teddington, three miles downstream.

Firemen later winched the sodden car ashore and were amazed to discover that once the water had flowed out, the car started first time! But Puddifoot told a local newspaper that unfortunately it never smelt the same again.

Police Constables Chris Howell and Roger Mercer, motorway patrol officers on the London orbital motorway, the M25, were sitting in a lay-by next to the slow lane when a car travelling in the fast lane suddenly stopped opposite them. 'We're lost,' shouted driver Peter Cray across the other two lanes as following cars braked heavily and began crashing into one another, eventually causing an 18 vehicle pile-up.

'Fortunately no-one was seriously hurt,' said PC Howell, 'but Cray gained the distinction of becoming one of the UK's most unpopular and most stupid of motorists all in the space of two minutes.' Cray was later charged with driving without due care and attention, and parking on a motorway, and was banned from driving for a year.

In America in January 1994, an FBI agent who was covertly tracking a suspect got himself into a very sticky situation. He was warned by workmen not to drive over fresh concrete but the intrepid secret agent ignored them – and ended up stuck in the new section of highway in Iowa.

Not only was his cover blown, it took 15 workmen and four tow trucks to get him and his car out of the concrete, and the FBI was charged $70,000 for the damage caused. Oh yes, and the bad guy got away.

When he was Governor of the Bank of England, The Right Hon. Robin Leigh-Pemberton was renowned as a man who rarely put a foot wrong. Well, perhaps not all the time. On a visit to the Wimbledon tennis championships in June 1993, Mr Leigh-Pemberton – he was later elevated to the aristocracy – leapt from his slowly-moving chauffeured Rolls-Royce as police told the driver to move on. One of the rear wheels of the two-tonne car then rolled over the Governor's right foot, leaving him hopping mad.

Later, he was asked to move along once more as he picked the wrong seat in the Royal Box. All told, a bit of a black Monday for the UK's top banker.

French drivers have one of the worst reputations in Europe, a fact illustrated once again in 1992 when despite the motorways being blocked for two weeks of the year by striking lorry drivers, there were still more than 9900 traffic deaths in France, an average of more than 27 every day.

As Sean MacCarthaigh wrote in *The Times* in August 1993, 'Behind the wheel the French have consistently proved themselves as the most reckless, bloody-minded and accident-prone in Europe, with almost twice as many fatal accidents as anywhere else.'

Now that may only be his opinion, but have you driven over there recently? Back in 1989 I was driving carefully along a fog-bound French motorway where visibility was down to a few feet, and yet reckless native drivers continued to zoom past me so quickly I hardly caught a glance at them before they were swallowed by the thick grey stuff.

Four or five miles up the road, I discovered what had become of most of them. In the outside lane there had been a fairly major accident, involving around 15 cars and vans. Thankfully, no-one was seriously hurt, but they had certainly been trying their hardest. I first suspected something was wrong when suddenly, looming out of the mist, I saw a crowd of gesticulating people standing in the fast lane – not gesticulating at traffic, as might have seemed reasonable, but rather arguing amongst themselves, presumably about who had managed to hit the accident at the highest speed! I managed to avoid them then, just as I avoid them now by not driving over there unless I absolutely have to!

Under the headline 'Vision of an Idiot', this story from Ottawa, Canada, appeared in the *Daily Star* in August 1993.

A woman who claimed she had X-ray vision blindfolded herself and drove down a city street. Lauren Rhoda, 37, ran over a pedestrian, breaking his toe, and knocked down a lamp-post before ploughing into three parked cars, one of which was a police vehicle. She was fined £400 and banned from driving for two years.

Veteran racing driver Jackie Stewart crashed into another car while demonstrating his advanced driving skills (!) on a crowded highway in Phoenix, Arizona in March 1990. 'I just kissed it,' said Stewart. He jumped out and told the other driver, 'Sorry, my fault.'

MOTORING SHORTS

Falling asleep at the wheel while driving is a constant worry for the weary motorist. But going to sleep behind the wheel in a stationary vehicle and almost drowning in the process is arguably just as foolish!

It happened to two men who dozed off in their car as they sat at the mouth of the Annan river in Dumfrieshire, Scotland, waiting for a signal to release 450 prize pigeons for a race across the Solway Firth. They woke up to find their car half submerged and the trailer housing their prize racing pigeons half under water. Unfortunately, those pigeons on the lower deck of the trailer were drowned, but the men managed to struggle ashore. The police agreed not to name the men to spare their blushes.

'You idiots,' shouted the Duke of Edinburgh just before he was thrown from his carriage during a racing competition at Windsor. The team driving the Royal carriage felt the rough edge of 72-year-old Prince Philip's tongue after his carriage careered through a water obstacle, throwing him to the ground. The Duke soon picked himself up and carried on, but the following day he probably had more than just a stiff upper-lip.

A pensioner died in a head-on smash after driving the wrong way down a motorway in May 1991. Police watched helplessly on video monitors as the 70-year-old driver turned onto the wrong side of the M5 at Bristol. Flashed warning signs were too late to stop the old man hitting a sports car travelling in the fast lane. The other driver survived but was treated for shock.

The driver of a sports car was giving a large woman a lift in 1969 and she had stowed her handbag behind her seat. Near the end of the journey she knelt on the seat, facing backwards, to pick up her bag. The sports car driver had to jam on the brakes in a hurry . . . and his passenger reversed sharply into the windscreen. She was undamaged, but the windscreen was a write-off.

MOTORING *SHORTS*

Film company secretary Susan Ornstein was somewhat shaken when her Mini demolished a traffic bollard in London's Hyde Park in 1968. But not half as shaken as when she received the bill for putting it back. The total came to £118 which was quite a lot of money back then, a fact underlined by Ms Ornstein, who said, 'I shall have to pay by instalments,' and who also added, 'And the best part of it is that the bollard has been knocked down again.'

The bill from the then Ministry of Public Building and Works was very carefully detailed. It read: 'Hyde Park sub-carriage road, repairs to Rutland Gate bollard. Two labourers, 12 hours each; one bricklayer, four hours; one bricklayer's mate, four hours; one electrician, four hours; one electrician's mate, four hours; one fitter, seven hours; one fitter's mate, seven hours. Total labour charge: £53. Materials: one new cast-iron bollard post, one new octagonal lantern, two 30-volt 60-watt lamps, one bushel of sand, 14lb of cement, one quart of paint. Total materials: £66. Deduct scrap metal of one bollard. Total: £118.

There are two things that may amaze readers today. First, that so many people were employed to erect this new bollard – today it would be two men at the most – and second, that the quality of the replacement was obviously so good. I went past Rutland Gate recently and discovered that one of the plastic bollards was leaning at rather a haphazard angle, clearly the recipient of a vehicle's attentions.

Mind you, one plus today is that the bollard is more likely to be damaged than the car. When Ms Ornstein hit the cast-iron bollard in 1968, she not only damaged the bollard itself – but her car was a write-off!

Cockney radio star Derek Jameson almost came a cropper when he and his wife travelled for 30 miles in a death-trap of a motor: the second-hand BMW's nearside back-wheel was only held on by one very loose nut. Jameson was snoozing in the passenger seat while his wife Ellen drove the recently-purchased used car on a

MOTORING *SHORTS*

trip to the coast, unaware that the wheel could fall off at any moment. Ellen put the strange rocking motion down to the fact that it was the first time she had driven the car, and so reasoned that she just wasn't used to it.

But eventually the rocking motion got so bad that thankfully, and in the nick of time, the couple stopped at a garage – and learned that they had been only seconds from disaster. They had bought the red BMW for £6000, as a second car, and it was delivered to their London home only the day before the trip. When Derek complained to the car dealers, they insisted that all was in order when the car was delivered and suggested that thieves might have tried to steal the wheel before being disturbed. Certainly the Jamesons were disturbed!

In a court case in Herrington, Kansas in 1968, the town's court had to decide who should be judged at fault if a car was hit by a train when negotiating a level-crossing. The local paper, the *Herrington Sun*, reported, 'The courts have held that in the case of an auto driver who neglects the utmost precaution at a railway crossing and is struck by a train, he is guilty of negligence and not entitled to recover.' That'll make you look both ways before you cross.

Since her fiancé had a habit of spouting poetry as they drove along, the young woman driver from Southsea was not too worried when, as she took a blind bend at high speed in 1968, he shouted out in alarm, 'Brake, brake, brake!'

'Even I know that one,' she said. 'On thy cold gray stones, O sea!'. Then she hit the back of a coal truck and wrote the car off. Injuries were there none.

'The first my client knew of the accident was when it occurred,' said an outraged solicitor of a motoring accident in Abergavenny, Monmouthshire, in November 1963.

Now here's a bizarre set of figures. Did you know that in 1926 there were a mere 1.7 million cars in Britain yet 4886 people died that year on our roads, while in 1992 with 25 million vehicles on the roads the death toll had dropped to 4273?

Of course, cynics might contend that traffic now has to travel so slowly on our roads, thanks to the roadworks and sheer volume of traffic, that it's unlikely anybody will ever get sufficient speed up to actually kill someone! The truth is, however, that road safety has improved and cars are more able to stop quickly and safely, as well as being better designed and therefore more able to withstand accidents without causing injury to driver and passengers. For its part, the Department Of Transport contends that, 'a major drive against speed in urban areas, where most accidents occur, is beginning to pay off.' Indeed. During 1993 the deaths on our roads dropped yet again, to the lowest level for 50 years, and in fact since the 1960s the number of deaths on our roads has been halved.

Going back to 1926, there were other reasons, of course, for the large number of accidents, including the fact that there was no driving test of any kind, so any fool could get behind the wheel. Neither was there a Highway Code – though a survey carried out by HMSO in 1993 worryingly revealed that 82 percent of motorists questioned could not identify a No Vehicles sign – and nor were zebra crossings, cats-eyes or double white lines in existence, so that the fools mentioned earlier often attempted to overtake when it patently wasn't safe to do so.

Bear in mind also that road signs, which had been designed for the much slower-moving horse and cart driver, were barely-visible small tablets at the sides of the road; there were no motorways and no maximum speed limit, though top speeds were realistically around 60mph because the roads were generally so narrow or in poor condition, or both; street lights were few and far between; and pedestrians were in much greater danger because there were few kerbs or even pathways on roads between towns.

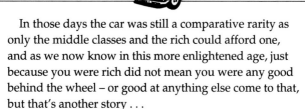

MOTORING SHORTS

In those days the car was still a comparative rarity as only the middle classes and the rich could afford one, and as we now know in this more enlightened age, just because you were rich did not mean you were any good behind the wheel – or good at anything else come to that, but that's another story . . .

When two motorists involved in an accident in 1968 near Driffield, Yorkshire came to exchange names, they refused to believe each other, for Mr Cyril White had been in collision with . . . Mr Cyril White! The two Mr Whites lived a mere 60 miles apart. They met in thick fog and though both cars were damaged, neither Mr White nor Mr White was injured.

We take the traffic light for granted these days, indeed at some busy road junctions some of us no doubt pray that they will soon fit one of the stop-go lights so that we can get out and be on our way. But the early traffic lights were not quite the docile things they are today.

In Britain the first proper traffic light did not appear until 1927, but in London a fairly simple manually-operated device was installed in 1868 so that Members of Parliament could drive into the House of Commons while the citizens waited – of course this was mostly aimed at horse-drawn traffic, the car not being common even in London until the early 1900s. Anyway, this first traffic light was fuelled by gas and involved the use of two revolving lanterns, one red, the other green. In 1869 the contraption unfortunately blew up, wounding the police officer who was operating it.

This one should perhaps go under the heading 'An Accident Waiting to Happen', but anyway . . . In Brazil in the late 1970s, criminals often tried to rob car drivers as they sat at red traffic lights. Because the police found it increasingly hard to catch the criminals in the act, drivers were told that it would be safest to ignore red

MOTORING  *SHORTS*

traffic lights at night – while taking a careful look in both directions first, of course.

Trouble was, this habit of ignoring the red lights soon spread to the daylight hours too! Rather belatedly, the Brazilian police started to re-enforce the law on traffic lights, but there are still some large towns in Brazil where red lights are routinely ignored.

MOTORING  *SHORTS*

ANIMALS IN CARS
Not a roaring success

Like animals and children in films, animals in cars are not always recommended, as some of the following found out.

Is it safe to drive if you have a donkey sitting alongside you? Good question, and one which a court at Totnes, South Devon, had to try and answer in 1969 when a man appeared before magistrates charged with not having proper control over his small Citroen 2CV. Peter Cox, then Principal of Dartington College of Arts, pleaded not guilty to the charge.

PC Kenneth Arthur told the court, 'As the vehicle drew near, I saw the backside and tail of a donkey through the windscreen and very close to the driver. On causing the defendant to stop, I confirmed that the animal was a donkey.' The officer added that the rear seat of the car had been removed and Cox's wife was sitting in the back holding the halter of the donkey, which was in a standing position. The front passenger seat had also been removed.

The prosecution contended that Cox did not have proper control because his visibility was restricted. Cox said the donkey was very tame. He produced a photograph showing that the driver could see between the top of the donkey and the roof of the car, and to either side of the car. While he was being cross-examined, the magistrates stopped the proceedings and dismissed the case.

All donkey-owning Citroen 2CV drivers could rest easy.

Landlord Bill Glover suspected something fishy was going on in 1967 when one of his customers came in and told him, 'I've just run over a shark in your car park.'

Bill's pub, the Spread Eagle in Norfolk Street, Reading, is 40 miles or more from the sea, so the landlord just laughed. But he soon stopped laughing when the customer returned, tugging a five-foot (dead) shark.

'It must have fallen from a lorry,' said Bill. 'The police were as baffled as I was – and they were not very keen to take it in as lost property. Now it seems I'm stuck with the shark, unless the owner comes to collect it.'

Anyone been to the Spread Eagle recently? Is the shark still there?

Mr and Mrs Ronald Cook of Rolla, Missouri, were on their way back from a squirrel-hunting trip in 1968 when they stopped their car to buy two baby pigs (as you do), which they put in the boot of their car along with their .22 hunting rifle. Suddenly a shot rang out. Twenty-year-old Mrs Cook cried, 'My God, it hurts,' before slumping down dead. One of the pigs had apparently bumped against the rifle and caused it to go off.

Brownie the mongrel really was in the doghouse in 1990 – after a crash course in driving! He jumped into owner Russell Leigh's £18,000 Mitsubishi Shogun, managed to release the handbrake, and rolled backwards into a 10-tonne truck at Stockport, near Manchester, causing £1500 worth of damage. No biscuits for Brownie.

In the mid-1960s, American car dealer Gordon Butler would accept almost anything as a trade-in, so it didn't come as too much of a surprise to those who knew him when he accepted an elephant as a down-payment on a new car.

It happened when an animal trainer from California walked into Butler's showrooms at Worcester, Massachusetts, and offered to trade in one of the three elephants he had with him in a trailer outside. After a phone call to find out the value of an elephant, Butler accepted it as a $1606 deposit on a Chevrolet. It was said that the car had a large V8 engine and a sizeable trunk –

MOTORING  *SHORTS*

just so the elephant trainer wouldn't miss his animal too much.

———

In 1987 the owner of a new Vauxhall Cavalier took the car back to his dealer in Rugby, Warwickshire, complaining of an intermittent squeaking noise from behind the dashboard. The garage mechanics were mystified until they removed the dashboard and found ... a hedgehog. It seems that somehow the animal had got in while the car was being built. Fortunately, it had survived and after being treated to a nice bowl of milk it was released back into the wilds.

MOTORING  *SHORTS*

THE PERILS OF PARKING

And the dreaded traffic wardens

Not surprisingly, this is one of the longest sections in the book, because just about everyone seems to have a horror story to tell . . .

You can't get booked for not buying a pay-and-display parking ticket if there are no machines to dole them out. That can only be the reason why someone chose to steal two of the machines – each worth £5000 – from a car park at Littlehampton, Sussex, in August 1989. The mystery was never solved.

Council officials in the pretty Sussex village of Ditchling hadn't reckoned on the anger of villagers when they started painting a no-parking yellow line through the village. First up was pop star Herbie Flowers, bass player with the supergroup Sky, who got out his own paint brush and started blacking out the yellow lines! He was soon joined by other villagers until there were as many as 100 people all furiously painting over the yellow lines.

It's really not very wise to throw several buckets of water over a traffic warden who's putting a ticket on your car, tempting though some might find it. For those who are not just tempted but actually carry out the heinous crime, the cost is £100. How do I know?

Well, because one motorist actually did it, and duly found himself in court in 1989 after throwing several bowls of water over traffic warden Lynne Williams. Later, he was fined by magistrates in Essex and when he came out of court, guess what? He had another parking ticket on his car.

MOTORING SHORTS

Offending the Queen Mother by parking a hearse by the side of a main road in London is just not on, at least that was the experience of Pete Townshend, star of rock group The Who, whose Packard Hearse disappeared from outside his rented flat in London's smart Belgravia just 12 hours after he'd bought it, back in 1964.

'I was 20 years old,' Townshend told a reporter from *The Sun* newspaper, 'and I borrowed £15 each from John Entwistle and Keith Moon to buy it. But the next day it had gone. I called the police, and they said it had been removed at the Queen Mother's request. Apparently she had driven past on the way from Clarence House to Buckingham Palace and objected to the hearse. I was furious that someone like the Queen Mother could, at the flick of a finger, have my car removed from the face of the earth.'

But the incident did have a bonus of sorts. It was apparently the inspiration for The Who's classic track, My Generation.

A woman who in one morning twice had a bust-up with a traffic warden nicknamed Eva Braun ended up spitting in her face. Ultimately she ended up in court, where magistrates ordered her to do 120 hours of community service and pay £25 compensation.

The unseemly scenes began when she shouted at warden Sarah German from her car because of a traffic jam, the court at Plymouth heard. Later, when Miss German saw the same yellow Datsun parked on double-yellow lines outside the motorist's flat in the centre of Plymouth, she was again showered with abuse and then with something else. The salivating motorist denied assault and inciting violence, but the court saw matters somewhat differently.

If you're a traffic warden in Southampton, maybe you should take a good look around before slapping a parking ticket on a car – and that includes looking up at the sky. Why? Well, because when a warden was about to put a ticket on the windscreen of a parked car in the town in March 1993, he was whacked on the head by a bottle hurled from a block of flats. The warden was not seriously injured, and the culprit was never brought to book.

Some local authorities in England have been encouraging home-owners to convert their front gardens into car-ports, such is the growing shortage of parking space in the cities. Indeed, it has been estimated by Government transport experts that all the major cities in Britain will run out of residential parking space within five years. Westminster Council in London has already introduced a scheme mating vehicles with properties and those cars not showing a parking permit are clamped.

If you're going to steal someone's parking space, it's best to find out first if it belongs to a lawyer – just in case. Because when someone parked in solicitor Tony Lawton's spot in York city centre, the lawman decided to take the matter very seriously indeed.

After 18 months' sleuthing, he found out who the culprit was and ended up winning a case against the parker, for trespass, collecting damages of £30.

'Hopefully it will act as a deterrent,' said Mr Lawton. 'It's better than clamping.'

The lawyer pays £370 a year for his parking space, which is just three minutes' walk from his office. 'I got fed up with going out to meetings in the middle of the day, coming back and finding that I could not park my car in a space for which I had paid,' said Mr Lawton, who added, 'The fact that I rent the space gives me legal title to bring a claim for trespass.'

The fined parker said, 'I've been too busy moving house, and didn't have time to defend the action. I've never admitted it.' The £30 damages were donated to a charity of his choice – Riding for the Disabled.

MOTORING SHORTS

A staggering 2500 parking meters are out of order every day in London, thanks to vandals and faulty machinery. And that adds up to a daily tally of 15,000 frustrated motorists clogging up the city as they drive around looking for somewhere legal to park.

However, there are plans to side-step the problem, one of the most promising being a scheme put forward by Movement For London. What happens is this: the motorist finds a meter which is not working and then he calls a special hotline phone number with details of the meter's position and his car's registration. He's then given a special number which he puts on his car's windscreen. The motorist can park legally, and the local council get to know where the faulty meters are: they can then rush around (this is what they said!) to get the meters working again.

Trouble is, the scheme is still under consideration. Let's hope they get it working properly soon; in the meantime I'll just keep driving around.

It seems you cannot even escape the attentions of the traffic warden when you're getting married, at least that was the experience of Britain's top horsewoman Ginny Leng. As she was in church marrying landowner Michael Elliot, a traffic warden was slapping a ticket on her car outside!

Farmer George Mellis was slung into a police cell overnight for not paying a £25 fine – on a lorry he didn't own! Innocent George of Oxfordshire was arrested on his own doorstep, and it wasn't until the next day that he was given the chance to explain that he'd sold the lorry a month before the fixed penalty fine was imposed.

Embarrassed magistrates had to apologise to George after checking with staff at the vehicle licensing centre, who confirmed that George had not been behind the wheel of this particular lorry for some time, and certainly not when the parking offence occurred.

MOTORING  SHORTS

The ticket had been slapped on George's old truck when it was parked in Slough in July 1990, and when the fine wasn't paid the Slough court passed the matter to Thames magistrates near George's home. Eventually, bailiffs were brought in and called on the startled George, who had not heard of the matter before. He explained the mix-up, but then out of the blue the court issued a warrant for his arrest.

Thankfully, the matter was eventually sorted out, and George was a free man once more.

Not content with introducing time-punch machines for priests last year, the Vatican City administrators decided in January 1994 to initiate the dreaded Denver Boot, or wheel clamp, because priests and nuns persist in parking where they shouldn't, causing regular traffic snarl-ups in the world's smallest state.

Does it ever seem rather strange that we place so much trust in the world's diplomats – the people who should be making the right decisions in our names, the ones who should be running the countries of the world safely, efficiently and smoothly? These are the same people who depressingly often disregard some of the most basic parking legislation!

We're talking not-paying parking fines, and yes, diplomats are the worst group of people in the world in that respect. If you are in any doubt, just take a look at some of the astronomically large arrears on parking fines run up by embassies both in the UK and in the US, and apparently in many other parts of the world. In May 1993 Turkish diplomats in London were singled out as the worst evaders, owing more than £100,000 in parking fines, just pipping the Russians, in second place, the French in third and the Nigerians, fourth. The Turks had totalled 245 unpaid parking fines in 1992, 115 higher than the year before, while the Russians failed to pay 226, down 137 on the previous year. The only bright spot was that persistent British pressure had paid off, albeit to a limited extent, in pushing the total of unpaid diplomatic parking fines down to 4166 in 1992, compared with 5145 in 1991.

Across the pond it was a similar story. In December 1993 it was revealed that Washington-based Russian diplomats owed $750,000, while other main offenders included the Israelis, Egyptians, Bulgarians, Nigerians and – surprise, surprise – the French, who owed $150,000 on 3762 tickets.

What about the Brits? Heartwarmingly, not a penny owed. What more can I say?

The mistake a salesman from Great Eccleston, near Blackpool, made was to ignore 15 parking meter tickets. His reticence in paying landed him with a whopping bill in May 1993 for no less than £9000, then the highest ever fine under the pay-as-you-earn fines system, which has since been radically modified. However, this was no relief to the driver at the time, who said, 'I can't believe it. I'd have got away with less if I had robbed a bank.' I think he meant less of a fine, not less money.

How much does it cost for a short line of yellow no-parking paint outside someone's house? Well, it should cost around £40, but in Waltham Forest in north-east London it cost £1600 – and it still hadn't been painted by the time the bill had come to that ridiculous amount.

Indeed, the local council's estimated costs for painting the line, outside the home of Guyana's High Commissioner, rose by a staggering four thousand percent! First, the request went to administrative staff in the council's planning department. It was then passed on and processed by officials in the highways department. After it was approved by them, it went to the works department, but not so that they could paint the line. Oh no, the works department just had to put up special signs notifying local people of the coming change. Then, an agency which deals with council advertising in the local press was brought in to help publicise the painted line. Finally, the council's legal department became involved because of regulations governing diplomats' parking spaces. 'The authority is just acting by the rules,' said a council spokesman in November 1993.

'This is happening just as the council are closing a home for the mentally disabled, to save money,' fumed one disillusioned councillor. After all this, it was revealed that the whole area was to be designated a controlled-parking zone in any case – cost: £50,000 – and that the High Commissioner's parking space would come within those regulations anyway!

The Commissioner, Mr Laleshwar Singh, regretted all the fuss: 'I thought it was just a question of marking the lines on the road.'

In November 1993, a driver who answered the call of nature by a road in East London turned around only seconds later to find a £60 parking ticket stuck on his windscreen and the traffic warden nowhere to be seen!

A driver from North Kensington, London, boobed when he made traffic warden Julie Spark *eat* the parking ticket she'd just slapped on his illegally-parked van! Julie got justice when the man appeared before Horseferry Road magistrates in October 1993, charged with common assault. The magistrates awarded Julie £150 compensation for her distress, and conditionally discharged the van driver for two years. That must have given him something to chew over.

Sue Daniels parked her Vauxhall in a London street near her home one night. When she came out in the morning she found that workmen had erected parking meters and her car had been clamped for parking illegally. 'The concrete around the meter was hardly dry,' wailed Sue, who complained to the local council. They eventually (grudgingly) refunded her clamping fee.

MOTORING SHORTS

The wheel-clamp company which towed away a Tory MP's car really got his goat – or rather his dog, because his West Highland terrier was inside the vehicle at the time! It seems to have backfired (not the car itself, but the clamping of it) because the MP, Anthony Steen, immediately launched a Deregulation Bill in the House of Commons aimed at dramatically cutting the number of rules and regulations governing the private towing and clamping companies.

'The more they take away, the more they earn,' said Mr Steen, 'and they rely on most victims neither having the energy, money nor determination to pursue them.' The MP had parked his car in a pay-and-display bay at Waterloo Place near Pall Mall, paid his money, put the ticket on his windscreen, and waved goodbye to his dog. When he came back out a short time later, both dog and car were gone. 'Five minutes after I'd left the car, I am told, the car was lifted off the road and taken to a pound. Apparently, one of the car's wheels may have intruded into the adjoining parking space.'

Bananarama, the female singing trio, have had their share of run-ins with the hated wheel-clamp, the most serious of which was the towing away of a car belonging to one of the band, Keren Woodward. 'My dog was still in the back,' said Keren. 'I couldn't believe they'd be so cruel.'

It seems that nowhere is sacred to the dreaded traffic warden, at least it seemed that way for bus driver John Lawes who was served with a parking ticket as he relieved himself in a public toilet in Brixham, Devon.

'I'd only stopped for a couple of seconds,' said John, 'when he came in behind me and told me to move the bus immediately, but by that time my mind was on other things and there was no turning back. I hardly had time to zip myself up before he thrust the ticket in my hand.' Traffic warden Mike Driscoll handed John a £20 fine for parking in a restricted zone.

MOTORING SHORTS

John Major was fined £60 by a court after revealing that he'd lent his car to Norman Lamont. Norman was the culprit, said Major, he was the one who parked my car on the yellow line.

But this is not quite all it seems. John Major is in fact one Danny Bamford, president of The Monster Raving Loony Party, and Norman is a colleague, a member of the same. Both of them had changed their names in a bid to get elected to Parliament. Mr Bamford had even gone so far as to rename his house at Wallington, Surrey, 10 Downing Street.

A young secretary's only mistake – but it turned out to be a bit of a major one – was not getting her post forwarded to her new address when she moved following her divorce. This lapse of memory led to her being arrested by police, handcuffed, put in a cold police cell, searched by female officers, her necklace removed in case she tried to commit suicide, and eventually being sent to court in a police van with male prisoners.

What was her serious crime? Believe it or not, non-payment of a £30 parking fine which she never even knew about! 'Nobody could tell me where my parking sin was committed,' she said. 'I have never been in hiding. I didn't know anyone was looking for me, and it took the police 30 minutes to trace me,' added the mother-of-two, who previously had no criminal record.

British actor Derek Nimmo was let-off a parking charge when the policeman due to give evidence against him failed to get out of bed in the morning. The star had been booked for double-parking his Rolls-Royce for 'two or three minutes' in Kensington, London. Mr Nimmo decided to fight the case at Horseferry Road magistrates' court but after a wait of nearly three hours the police officer had still failed to turn-up to give evidence.

'The officer is on a late-turn,' said prosecutor Vivien Walters, 'and it is anticipated that he is in bed.' The case was dismissed.

MOTORING *SHORTS*

Beware if you leave your car at London's Heathrow airport. In November 1993 one of the computers went haywire and billed each driver £64,000 for 15 years' parking! Of course, drivers couldn't get their cars out of the car park until they paid up. Just what you need when you come back from holiday – who said computers never lie? The car park company now employs clerks to routinely check the machines' demands for cash.

A British motorist who dodged a 50p parking fee was fined an astonishing total of £700 because he did not return a form asking who owned the vehicle. The Liverpool man pleaded guilty to the court by post and he later described the fine as 'ludicrous'.

In Britain, the dreaded Denver Boot was introduced in London in 1983, but at first it did not succeed in trapping all of its intended victims. Police were so unused to working the clamp that a Ford Granada they were trying to attach it to drove off as a police officer down on his knees fiddled with the clamp's bolts.

In another instance the driver was more imaginative, but less fortunate. He was a Volvo owner who jacked his car up, managed to undo the bolts on his wheel, put the spare on and was about to drive away . . . when the police turned up, blocking his getaway. The motorist was fined an extra £200 for tampering with the clamp, on top of his parking fine and clamp-removal cost – though this last fine seemed particularly unfair, considering the chap had already removed the clamp without any help from the police.

Piero Maestrini was saved from death by a policeman who dived into a canal in Lodi, Italy to pull Piero from his crashed car. However, Piero did not entirely escape unscathed – the policeman promptly wrote out a ticket for illegal parking and presented it to the hapless Signor Maestrini.

MOTORING *SHORTS*

As a 68-year-old volunteer delivered meals-on-wheels to old and infirm people in Bournemouth, Dorset, her car was wheel-clamped and she had to pay a £50 fine before she could continue to deliver the by then decidedly lukewarm meals. Sylvia Wells, meals-on-wheels chief for the area, branded the private clamping company as 'hard-hearted'. They later pledged to return the cheque – but didn't offer to re-heat the meals.

A van driver in the diminutive Isles of Scilly, just off the southernmost tip of Cornwall, became the first ever islander to receive a parking ticket in November 1993 when he parked on double-yellow lines on the island of St Mary's. The islands have had the same rules and regulations as mainland Britain for just as long, but until recently the two-man police force had never bothered to enforce them. A crackdown was thought necessary to reduce increasing traffic congestion on the island, which measures about five miles by four.

A grieving widow was asked to pay her husband's parking ticket, even though he died the day after being booked. The woman, who lived in Devon, said, 'I couldn't believe how heartless these people could be. My husband had gone to the doctor's with severe chest pains, and he died the next day. His mind was elsewhere at the time.'

A spokesman for Mid Devon Council said, 'We get so many letters it's difficult to pick out the genuine ones. But her husband was alive when he got the ticket.' Later they agreed to a refund.

It's not a good idea to throw a parking ticket on the ground, at least that's the experience of a motorist in Kent who was fined a massive £1000 after he'd removed the ticket from his Jaguar and hurled it on the ground in disgust. He was fined for dropping litter!

MOTORING SHORTS

Everytime Queen Elizabeth II goes to Sadler's Wells ballet, Raoul Gonzalez has to fork out £100. It always happens because while the Queen and other VIPs get the red-carpet treatment, Raoul's car is towed away by security-conscious police. Even though Raoul has a valid parking permit, for which he pays Islington Council £120 a year, and he hasn't parked illegally, he has to pay the £100 to get his car back from the police pound. Islington's chief parking officer suggested that Raoul and his neighbours watched their cars 24 hours a day!

Traffic warden Brenda Bussey booked her own husband for parking on yellow lines when she failed to recognise the family saloon car. Husband Stan had just popped into a shop to get some goods when Brenda slapped the parking ticket on the car and then moved on. She only realised what she'd done when she returned home to see Stan waving the ticket at her. 'Yes, I have forgiven her,' said Stan sportingly.

When armed SWAT police swarmed around a Los Angeles home, the fugitive slipped outside and phoned the police to say he was giving himself up. But as the cops nicked him, they found he wasn't the violent mobster they were after – instead he was only wanted for not paying a couple of parking fines.

If your car breaks down, try and ensure that it's not on double-yellow lines – that's most probably the advice of a Southsea man who was fined a massive £500 when he left his expired vehicle for 20 minutes as he went to a garage for help. Put another way, his parking fine cost him an amazing £25 a minute! The ironic thing is, if the driver had been given a fixed-penalty fine it would have cost him £36, but because he believed he was in the right he decided to take the case to court. But Portsmouth magistrates found him guilty and imposed the massive fine.

MOTORING  SHORTS

Taxi drivers in Tiverton, Devon, gave tough traffic warden Terry Forward – nicknamed Hitler – a 52nd birthday treat by deliberately parking their cars on double yellow lines so that Terry could book them!

British TV star Bella Emberg was more than surprised to receive parking tickets for places she had never visited. The other surprise was that the perpetrator who kept getting booked was a man, who drove the same make of car, in the same colour, and with the same number plates! At one point Bella had five parking tickets which she had not personally incurred.

When she received a summons for allegedly failing to stop for the police she decided enough was enough. Her solicitor proved she was the victim of a bizarre mix-up by showing that the reference number on her Mazda tax disc didn't match the one printed on the parking tickets.

An elephant called Nelly had a £40 parking ticket stuck to her side while she waited for her owner on yellow lines outside a bar in Naples, Italy in March 1993. Apparently he was inside making a trunk call.

A policeman asked a funeral director to interrupt a service for two nuns so mourners could move their cars. When the funeral director refused, the cop slammed £20 tickets on two vehicles, one owned by a nun!

'The officer's attitude was unbelievable,' complained the funeral director, Mr Hyland, 'it really annoyed me. All that was being said could be heard inside the church.' An East Sussex police spokesman said the cars were obstructing a busy road outside St Bartholomew's Church.

Cops slapped a parking ticket on a donkey after gold prospector Jackass Larry left her tied up outside a cafe at Grass Valley, California on 13 April 1993.

MOTORING SHORTS

Dentist Walther Hanning was fined £14,500 after being given a parking ticket in Bayreuth, Germany. He ripped up the £12 ticket, parked his Porsche on the traffic warden's foot, then drove at him three times, injuring his legs. Interesting. Wonder how business is at the dentist's surgery at the moment?

This one's a story of bungling police tow-away squads. Michael Hughes of Brighton, Sussex, reported a car that had been illegally parked outside his house for a month. After waiting four weeks, he again telephoned the traffic wardens. Still no response. Another week went by before the tow-away squad turned up at his home – and carted *his* car off to the pound.

'I couldn't believe it,' said an exasperated Michael. 'When they did get their act together they towed away my car which was illegally parked for a matter of minutes, while the other one had been in the two-hour waiting zone for months.' It cost Mr Hughes £100 to get his car back.

A firm in Cardiff, Wales, ordered its workers not to park near the National Stadium because of the World Cup qualifying match that evening between Wales and Belgium. All well and good, but the match was being played in Brussels!

A mix-up by Dumfries and Galloway Regional Council in Scotland meant that thousands of fixed-penalty parking fines had been issued by mistake and were not legal. The problem arose because though the fines appeared to have been issued quite correctly, the council had not had the correct statutory authority signed – or so it was claimed by solicitor Alistair Oak in August 1989. It meant that over a period of four years, motorists had technically not been liable for the fines, and they were therefore advised to claim back any money that they had paid.

Tom Field of Turnditch, Derbyshire, was somewhat mystified when he received a parking ticket for leaving a steam-roller he doesn't own in a city he's never visited. A spokesman for Edinburgh City Council in Scotland admitted that 'somewhere along the line a mistake may have been made'.

Christopher Roe of Seaford, East Sussex, thought he'd be able to forget about the £26 parking ticket he got on a visit to Philadelphia, USA. No such luck, though he almost got away with it when bungling American cops sent the ticket, for some unknown reason, to Bogota in Colombia. However, after the ticket's 2000-mile jaunt to South America it eventually arrived at Post Office HQ in London where clever detective work discovered that Sutton Drove, Bogota, East Sussex, was in fact Seaford, East Sussex. 'The Post Office did well,' said Christopher, 'but I wish they hadn't bothered.'

A blunder by a council officer at Brighton in 1993 meant that hundreds of parking tickets slapped on motorists' windscreens were in fact illegal. The major mistake came to light when a junior legal assistant was dismissed after paperwork from 23 unfinished road traffic orders was found in the bottom draw of his desk at the Council offices. It meant that some parking restrictions had been put into force before they were signed by the responsible senior officer, and that meant that the traffic orders were being illegally enforced. Join the queue of disgruntled motorists.

If you're going to park your car in Maastricht, it's wise to find out first if you just happen to be parking outside the home of a four-star Nato general. If you're not sure what might happen, have a word with Britons Marion Silverstone and son Marc. Their red Citroen BX was blown apart by anti-terrorist police in the Dutch town after they sent robots in to investigate the parked car, evacuated nearby houses, and then carried out two controlled explosions which blew the doors of the car off and ripped up the floor.

The police admitted they were shocked when their two 'terrorist suspects' turned up. A spokesman said, 'We apologised and I think they laughed in a small way.' A very small way, apparently. The car was later repaired for the Silverstones at the authorities' expense.

One of the most potentially dangerous British Government positions is that of Minister of State for Northern Ireland, and its incumbents always have to be very much on their security guard. So it was rather surprising that former NI Minister Michael Mates didn't seem to realise the near-panic that he would cause when he abandoned a car belonging to a friend of his on double yellow lines smack outside the Houses of Parliament. Fortunately, of course, the car did not go off with a bang, and Mates later apologised for the blunder which caused absolute traffic chaos around Westminster, and which cost the taxpayer some £20,000 in the process as the security forces swung into action.

Chelsea football fans who regularly park across the driveway of Conservative MP Rupert Allason risk being clamped by him. 'Offenders could wait hours until I return from the House of Commons,' he said. But Allason would be taking a real risk if he carried out his threat. According to neighbours speaking in February 1994, there was no parking problem, and a spokesman for Chelsea Council warned, 'He has no right to clamp cars in the street.' One-nil to Chelsea.

Furious Gary Frayne was locked up in a cell for not paying a parking fine – but he doesn't even have a car! Gary was taken to court in February 1994 despite protesting to police that he had got rid of his Ford Cortina some four years previously. Fortunately, it took

MOTORING *SHORTS*

magistrates in Northumbria, England, just five minutes to realise that a mistake had been made, and Gary was freed.

His nightmare began when court official Dennis Hogg knocked on the door of his home with a warrant for his arrest. Gary was marched to a police car and taken to the police station, all because he had supposedly not paid a £30 parking fine. 'They took my liberty away for three hours for something I didn't do,' he wailed.

Northumbria police apologised for the mistake, and a spokesman said, 'Clearly, someone deliberately gave Gary's name when the offence occurred.'

When Princess Diana and pal Lucia Flecha di Lima went shopping in New York's top stores in January 1994, they made a whole series of parking blunders – getting $175 worth of tickets in all of the swankiest of streets, despite the pleas of the Princess's British security men to New York's policemen. The traffic cops in Fifth Avenue were simply not impressed by the Royal credentials, though the shops were – Di and her friend spent a fortune before retiring to their $1000-a-night hotel.

Housewife Alex Laughlin was overjoyed when she won a new Austin Metro in a draw in January 1990. But when the mother-of-two from Eastleigh, Hampshire, collected her prize from a Bath shopping centre, she found a parking ticket already on the windscreen!

A traffic warden had a trick up his sleeve for TV magician Paul Daniels in March 1994 – he slapped a £30 parking ticket on his Bentley as the star recovered in hospital from a food poisoning episode.

Heartless or what? No mate, just doing my job.

MOTORING *SHORTS*

THE LONGEST JOURNEY

Setting out on a motoring trip is not usually daunting, but for one driver it soon turned out that way . . .

A furious wife who had a less than perfect sense of direction drove off to confront her husband's mistress – and got lost on an amazing 1350-mile round-Britain trip. She arrived back home exhausted, 27 hours later, without finding her love rival.

The mother-of-two had angrily set off on her mystery tour at three o'clock one afternoon after her husband confessed to having an affair while working away from home. She jumped in the car and motored the 250 miles from her Norwich home to Middlesbrough, where the other woman lived, knowing only the woman's first name and that she visited a local Labour club. The distraught wife toured the area in tears for four hours but didn't find the woman, and so eventually decided to give up her hunt and return home.

Which was where the fun really started because she couldn't find the main A1 trunk road. Police who were breath-testing a motorist told her the way – but she became confused and headed north instead of south. It was 2.30am before she realised she was going the wrong way, but by then she was near Durham in the north of the county. She used an emergency phone near Durham to ask for directions, and at long last started to drive south, back towards her home. All seemed to be going well until she suddenly found herself back at the very same phone box she'd used earlier – she had gone all the way around a roundabout and headed back north again! Police again tried to point her in the right direction but to no avail: at 6am she arrived on the outskirts of Edinburgh in Scotland.

She didn't even realise she was in Scotland until she heard the accents of two traffic policemen, who tried to direct her back down south. In a desperate effort to get her on the right track, the officers actually escorted her back to the A1 and pointed her south. At last, she was going in the right direction, but it really wasn't her day . . . she ran out of fuel, and having had to refill her Renault's petrol-tank three times on her mammoth trip at a cost of £80, she had no money left. She broke down in Wetherby, West Yorkshire, and from there phoned her mother, who came out to collect her.

They eventually got her back home at 6pm, the day after she left, and she said of her husband, 'I still love him, despite what he has done, and want him back.' Let's hope he's got a slightly better sense of direction, otherwise they'll never get back together again; not unless they meet on a motorway somewhere in the north . . . or the south . . . or . . .

MOTORING  *SHORTS*

AMAZING CLAIMS

Insurance pleas with a difference

Heard the one about the man who looked at his mother-in-law and then immediately drove his car down over an embankment? Or the man who wrote-off his car because he drove into a telephone pole as he tried to squash a fly on his windscreen?

These are just a couple of the many amazing and amusing claims received every year by the country's insurance companies. For every thousand or so insurance claims which are straightforward, there is one which is anything but. Some are entertaining simply because of the way they have been written up, while others are the result of genuinely bizarre circumstances. But all of them are truly amazing claims.

I was on the main road and the car in front was turning right. My attention was diverted from the road to the footpath by a scantily clad, bra-less girl. When I looked back it was too late to miss the car.

I was admiring a young lady walking along the opposite pavement. It's a habit. I can't help it.

I had one eye on a parked car, another on approaching lorries, and another on the woman behind.

Whilst riding in a car my spectacles were blown out of the window. I was leaning partly out of the window at the time, attempting to get some fresh air, on account of

---

These claims have been supplied by some of the UK's largest insurance companies, including General Accident, Norwich Union, AA Insurance and Hogg Business Rental's insurance division, to whom I am most grateful.

MOTORING  *SHORTS*

being a little bit drunk. I have not lost my glasses by leaning out of car windows before, but does my insurance policy cover drunkards doing it?

Whilst operating a combine harvester and discharging grain into a trailer drawn by a tractor travelling alongside me, I found it imperative to inform the driver of the tractor that he was (and is) a pillock.

This badly chosen word, delivered with the force applicable to the circumstances and at a volume required to render it audible to the said driver, resulted in my temporary upper denture being projected into a pile of straw some twenty feet away. A search was made at the time, the place marked, and another search was made the following morning, both without success.

Driving along at 45mph, on a straight road, when the passenger coughed and lost his false teeth. On trying to recover them, he bent down and accidentally pulled the steering wheel, therefore causing the driver to lose control and crash.

The third party sneezed and his false teeth fell out. While looking for them he crashed into the back of the other car.

After a motor accident in the South West, a claimant laid the blame fully on a dog which had run into the road and then, 'ran away without stopping to exchange names and addresses as required by law'.

I pulled into the lay-by with smoke coming from under the bonnet. I realised the car was on fire so I took my dog and smothered it with a blanket.

I knew that the dog was obsessive about the car but obviously I would not have asked her to drive it if I had thought there was any risk.

Whilst proceeding through Monkey Jungle my vehicle was enveloped by Small Fat Brown Grinning Monkeys. Number 3 Small Fat Brown Monkey (with buck teeth) proceeded to swing in an anti-clockwise direction on my car's radio aerial. Repeated requests to desist were ignored. Approximately 2 minutes 43 seconds later, Small Fat Brown Monkey disappeared into Monkey Jungle clutching my radio aerial.

The horse bit the soft top roof of my MG sports car, tearing the material.

A Hofmeister bear got off the back of the lorry. The lorry driver got out and one of them said to the car driver, 'Sorry mate. It was our fault.'

Nature of damage: 'Broken windscreen'. Persons injured: 'A French owl'. Injury sustained: 'Headache'.

First car stopped suddenly, second car hit first car, and a haggis ran into rear of second car.

In an attempt to kill a fly, I drove into a telephone pole.

Please delete Mr Smith from the policy with immediate effect, as he has been unable to start the vehicle.

I had just moved off in first gear and then into second, when I found I could not turn the steering wheel and so I applied the footbrake, but there was no response. Then I discovered that I had forgotten to take off the Krook-Lok from the brake pedal. I had barely travelled 20 yards and I hit a parked car.

I would wonder how your insured could be in full control of car while eating large-size bread cob, which he continued to eat even after the incident!

Miss X was not guilty. The pedestrian knocked himself down by running in front of my car when we were moving forward slowly.

The injured party ran across the road in an attempt to commit suicide.

The pedestrian had no idea which way to run, so I ran over him.

A pedestrian hit me and went under my car.

To avoid hitting the bumper of the car in front I struck a pedestrian.

I was sure that the old fellow would never make it to the other side of the road when I struck him.

I saw a slow moving, sad-faced old gentleman as he bounced off the roof of my car.

I knocked the man over. He admitted it was his fault as he had been knocked over before.

I misjudged a lady crossing the road.

Q. Where is the building situated?
A. On the side of the road.

Q. Whose fault was the accident?
A. I cannot really say – my eyes were shut at the time.

Q. Could either driver have done anything to avoid the accident?
A. Travelled by bus.

Q. Purpose of journey?
A. Dirty night out.

MOTORING *SHORTS*

Q. Was any person injured?
A. I don't know. They were taken away in an ambulance.

Mr X is in hospital for an operation and says I can use his car and take his wife while he is there. What shall I do about it?

A large branch was ripped off and fell across the car. The tree is now noticeably damaged and disfigured.

It would be useless for me to try and assess his speed. Any vehicle which is obviously going to hit me is travelling too fast.

I parked my car outside a Chinese take-away restaurant and when I returned it was gone.

No witnesses would admit having seen the mishap until after it happened.

It is particularly unfortunate for me to have this charge of careless driving brought against me as it is the only thing I can do really well.

My car was broken into whilst parked in my driveway. As the car was parked on my property at the time I wish to make a claim on my contents insurance policy.

Whilst travelling along, an item fell from the passenger seat and I reached down to pick it up. Unfortunately I took the steering wheel with me and caused my vehicle to collide with the third party's.

Whilst attempting to demonstrate the vehicle's capabilities over a fairly rough terrain, unfortunately it hit a pothole. I lost control, the vehicle spun around and hit a tree at its nearside.

MOTORING  SHORTS

Vehicle caught fire whilst in motion outside Calvert Lane fire station. We drove in and they put fire out.

Third party who had just failed driving test was asked to park vehicle. While doing so she ran into the stationary vehicle.

He was a front-seat passenger in a mini bus. As a result of this accident the lower part of his body was held in the vehicle and the other part of his body thrown through a side door, though, needless to say, both parts remained connected.

I parked the car and I was sleeping inside. When I got up found the car was missing.

The car came out of the gate, failed to stop and ran into the side of my car (the road was dry after rain).

If you think I'm a borderline case, whether to repair me or not, can I tell some of my good points? I've a new (not reconditioned) engine, new front wings and a vinyl roof, and gas-filled shocks fitted last year. Inside, I'm pampered with a wooden facia and glove box lid, hazard warning, cigar lighter, rear-window demister, extra instruments, radio and stereo tape-deck and Fiamm air horns. Since Xmas I've had electronic ignition fitted and a new near-side sill and new battery. To Mr and Mrs Dolphin – I'm great!

The back axle broke and tyre rubbed against body and caught fire. Mr G was driving a car and he went off and phoned the fire brigade and found a fire extinguisher whilst I kept the flames under control by filling my wellington boots with water from a cattle trough in a field.

MOTORING SHORTS

Many thanks for your help. Will you now please give the Hillman Hunter a good funeral?

I had been shopping for plants all day and I was on my way home. As I reached an inter-section, a hedge sprang up, obscuring my vision and I did not see the other car.

I was doing 28mph. I am positive of this as I was looking at the speedo when I hit him.

A motorist from Cheadle, Lancashire seemed somewhat confused by the accident he was involved in. 'I was driving in slow moving traffic when a learner driver lost control of his parked vehicle and ran into the back of my car.'

Coming home I drove into the wrong house and collided with a tree I don't have.

The other car collided with mine without giving warning of its intention.

I collided with a stationary truck coming the other way.

A truck backed through my windshield into my wife's face.

I pulled away from the side of the road, glanced at my mother-in-law and headed over the embankment.

I had been driving for 40 years when I fell asleep at the wheel and had an accident.

I was on my way to the doctor with rear-end trouble when my universal joint gave way causing me to have an accident.

MOTORING  SHORTS

As I approached the inter-section, a sign suddenly appeared in a place where no stop sign had ever appeared before. I was unable to stop in time to avoid the accident.

My car was legally parked as it backed into the other vehicle.

An invisible car came out of nowhere, struck my car and vanished.

I told the police that I was not injured, but on removing my hat I found that I have a fractured skull.

The indirect cause of the accident was a little guy in a small car with a big mouth.

I was thrown from my car as it left the road. I was later found in a ditch by some stray cows.

The telephone pole was approaching, I was attempting to swerve out of the way when I struck the front end.

In Plymouth a motorist made the following statement: 'With regard to blame, I can only say that if there had been a pavement on the side, and the cyclists had been on it instead of the road, then I would have probably missed them.'

A London driver sent in this rather mundane claim which stated merely that, 'The car was left at 6pm. I came back at 10pm and it was gone.' But the accompanying sketch was the laugh. The sketch plan showed the street where the car was parked. The Before section showed the car parked by the side of the road, the After box showed a blank space where the car had been.

MOTORING *SHORTS*

An accident report came in from a motorist who had been cruising along a motorway on his own. He wanted to get something from the glove compartment as he went along, but found it was locked. So he took the keys from the steering column to unlock it as he was driving. His steering locked and he ploughed into the crash barrier.

I reversed from the road onto a shopping frontage car park and did not see the lamp post coming out of a shop rear-end make contact.

A motorist from Portsmouth sent in this claim: 'Suddenly I saw a parked car and wondered if it was moving or not. I realised it wasn't and went straight into the back of it.'

A driver in Monmouth penned this claim: 'The other driver was to blame for driving in an erotic manner.'

A motorist from Stoke-on-Trent seemed to have had a similar experience: 'Whilst travelling along the A487 a FSO car travelling in the same direction slowed down in an erotic way, giving no warning by its brake lights or any reason, ie turning.'

Rumour has it that motorists in the Stoke-on-Trent area are now out-and-about most nights looking for the erotically driven FSO.

A policyholder filled in an accident claim after running into the back of a bus at a set of traffic lights. Fortunately, he had a witness. Unfortunately, the witness was not likely to be much use.

Why not? Let's let the policyholder tell us in his own words: 'He cannot read or write, and he is blind and deaf.'

MOTORING *SHORTS*

In a letter answering the question, 'Is your insured being charged with any motoring offence in connection with the accident?', the third party's solicitor declared, 'Our insured's driver is being executed in connection with the accident.'

A Glasgow claimant described his personal accident.
Q: How did it happen?
A: Removing engine from car. Engine fell and jammed my hand against chassis.
Q: What were you doing at the time?
A: Yelling like mad!

A Wolverhampton motorist was manoeuvering on a garage forecourt. 'I was watching over my left shoulder for parked cars and caught my offside front wing on a sign.' The wording on the sign was, Free Estimates For Accident Repairs.

Now, that was convenient, wasn't it?

When a Christian Centre Minister made his first claim for a motoring accident with Norwich Union, it was probably appropriate that he should have been involved in a collision with a van belonging to the Jolly Friar company.

I thought the window was down but it was up, as I found out when I put my head through it.

The witness gave his occupation as a gentleman but it would be rather more correct to call him a garage proprietor.

I was reversing in a parking space in a cuddle sac.

MOTORING  *SHORTS*

Rosemarie Spiers, a Director at Hogg Rental Business Services division, takes us through some of the more bizarre claims they received in 1993.

'Normally we expect the driver of the rented vehicle to complete the accident report form. Unless he's dead or out of circulation, of course. So we always take careful note when we receive one from the owner of a rental company. This particular one said, "Our hirer and a friend decided to hold-up an off-licence and use our vehicle as a getaway car. I understand on making his getaway he drove through a road block, damaging a police car and a civilian car, ending up parking our vehicle in a brick wall. Much to my surprise he still managed to escape the police."

'Trouble was, that was only half the story. The civilian car he'd mentioned – a beautiful new executive model – was owned by a real hero who witnessed the robbery, contacted the police, and then donated his car to help form the roadblock. Unfortunately, the renter rammed the hero's car five or six times, before abandoning the rented car and running off. Unfortunately for us, we also insured the hero's car! Now, to try and minimise the impact (no pun intended!) on the claims record of the rental company, we sought to recover the costs from the robber. Unfortunately, he was sent down for seven years, and, as armed robbery was more important than the lesser charge of criminal damages, we were put well down on the list of considerations.

'I wonder what the stated purpose of hire was?'

Another rather strange claim is described by Rosemarie Spiers. 'Some of our customers have diversified into other businesses. But there are, we discovered, some unexpected hazards to combining two very different enterprises under the same roof. One customer operated a very successful rental company alongside an equally well-run driving school. Now, on this particular day, this company had run short of rental cars, and the solution on this occasion was to rent out one of the driving-school vehicles.

'Unfortunately, in the rush, no-one actually disconnected the dual control system. At some stage, while the wife was driving, the husband panicked and slammed the brakes on. The action caught his wife by surprise and caused her a neck injury. Naturally, we said that, as the driver of the car, she would be unable to claim for her injury as, in theory, she would be claiming against herself. "I wasn't in control of the car," she claimed, "It was my husband who braked, not me!" '

Finally, one claim received by Hogg in 1993 was from a perfect gentleman who, on a clear sunny day in Bath, reversed his rental vehicle, 'scraping the side of a sensibly parked Toyota'.

MOTORING  *SHORTS*

COURT IN THE ACT

I put it to you . . .

It's an unfortunate fact that the antics of some drivers inevitably land them in court. Here are some of the more bizarre and often amusing cases of recent years.

A motorist sped past a carload of policemen at 100mph while shaving with his electric razor, a court was told in March 1990. The 22-year-old salesman from Beaconsfield roared up behind the officers on the M40 motorway as they were travelling to a Home Office conference. The salesman flashed his lights at them to try and make them move out of the way, then overtook them at between 95 and 105mph on the inside, and angrily waved his razor in the air at them as he went past.

Chief Supt James Carrs, who was in the unmarked police car, told the court, 'We couldn't believe it. When we indicated at him to stop, he looked over and gave us a V-sign.' The officers were so incensed with the man's driving that one of them showed his police warrant card through the car window. When the salesman realised they were policemen, he looked at them wide-eyed then suddenly veered off the motorway and up an exit road.

In court he denied reckless driving and speeding, but was found guilty. He said, 'I certainly was having a shave, but I had my left hand on the wheel. I don't remember giving anyone a V-sign.' He was banned from driving for seven months and fined £500.

A man who held a grudge against, 'fat b**tards who play golf' ended up in jail after driving his Austin Montego saloon car across a hotel golf course at high speed, terrifying golfers and ruining the greens as he skidded the car across the grass.

The driver launched his assault on the Manor House Hotel course at Moretonhampstead, Devon, because he was angry with the owners, whom he believed owed him money for window-cleaning, Exeter Crown Court was told in September 1993. Mr Richard Merrett, prosecuting, said the man caused £9000 worth of damage when he left the road leading through the course to the hotel, damaging the 13th and 14th greens.

His two women passengers were terrified. He had told them he intended to drive into the swimming pool, the court heard. But instead he drove back onto the course, narrowly missing one player, before executing a perfect 360-degree handbrake turn on the fourth green which left it unusable. Ignoring the shouts of outraged golfers, he then tore off the course again and onto the public road where he was eventually caught after a chase involving no less than seven police cars and a police helicopter.

Miss Corinne Searle, defending, said, 'It is a strange case. He can remember very little of what happened.' The man pleaded guilty to two charges of dangerous driving, three of causing criminal damage and one of driving while over the legal limit of alcohol. He was jailed for eight months, was banned from driving for two years and fined £500. Police Constable Ian Walker said the driver 'was down in the dumps over his girlfriend and his mortgage. He said he hated fat b**tards who played golf. But it was a good advert for a Montego.'

Really? I suspect it's likely to be some time before Rover advertises the Montego with a line like, 'Driven by window-cleaners who hate fat bastards who play golf.'

Scottish eccentric David Robertson didn't think he should pay fines for motoring offences because he'd crowned himself King of the Picts – the ancient and wild race which many centuries ago ruled Scotland. Because of his kingly title, Robertson said, he had renounced his British citizenship and was not subject to British road taxes because he had no agreement with the English crown.

Unfortunately for the plucky Scotsman, the Sheriff's court at Tain, Scotland didn't agree and he was found guilty of driving without road tax, a test certificate, or insurance. Sheriff Donald Booker-Milburn fined him £50 and ordered him to pay back-duty tax at £5 a week. Robertson claimed that some weeks he didn't even earn £5 and that being King of his tribe he didn't accept British state benefits or welfare. 'I borrowed a friend's car to get the groceries,' he said. The Sheriff told him to pay up.

A 70-year-old Englishman carried on motoring despite receiving a two-year driving ban from a Scottish court because he claimed he hadn't understood the Sheriff's Scottish accent and didn't realise he'd been banned. After hearing the case, the court in Stirling banned him for another year. Apparently he heard this time.

Police in South Africa don't see much point stopping a certain 21-year-old man from Port Elizabeth in connection with any motoring offences. He's one of three identical triplets, and a charge of driving without a licence was dropped when the prosecution admitted that positive identification would simply prove too difficult.

In 1993, a pensioner driving near Yarmouth was prosecuted by police because he was driving so slowly that he had caused a 25-mile traffic jam behind him.

In March 1994 a motorist from Cambridgeshire was allowed bail after being charged with wiring up his car to give an electric shock when he parked it in West London.

A young crook convicted of 74 car offences was sentenced by a court in November 1993 to ... free driving lessons, at the expense of the taxpayer! As one Conservative MP put it, 'We'll be giving cat burglars free tuition in abseiling next.'

MOTORING SHORTS

The youth's motoring offences included driving without a licence, no insurance and no roadworthiness certificate. He was seen by police in Bedford on three separate occasions driving high-powered cars. Fumed the *Daily Star*, in an angry editorial, 'It's a wonder the magistrates didn't give the poor, misunderstood wretch a new car.'

A car owner from Prestwich, Manchester, really should have taken the trouble to turn up at court to answer a string of mostly minor charges relating to his 20-year-old Ford Escort. The car, which was later sold to a scrapyard for £30 in March 1993, landed its owner with fines totalling a staggering £6000! At the time of the fine, the man was drawing £32 a week unemployment benefit.

'I can't believe this,' he said, 'they must think I'm a millionaire.' The fines were made because he failed to turn up at court, so each fine was worked out on the basis of an assumption that he had £100 a week disposable income.

Bruce James should never have decided to tour the new police computer centre in New York, because officers asked for his details to show other members of the tour party how well the system worked. The machines did their business and the boys in blue couldn't believe their luck: Bruce was wanted for jumping bail on a driving charge. He was later to be found helping the police with their enquiries.

Being world famous can have its problems, but trying to take pictures of the world famous can have its dicey moments too; at least that was the case outside singer Cher's home in Beverly Hills in 1989 when her then boyfriend Rob Camilletti drove his Ferrari at a photographer who was waiting to take pictures of the star outside her Hollywood residence. In the event, the photographer apparently got some unique pictures of a

Ferrari, the like of which few other photographers have managed. Mind you, he nearly got severely injured for his pains. Camilletti was fined $300 in Los Angeles on two charges of vandalism and told to do 300 hours of community service.

Here's an interesting one for crime-watchers. Take a young hooligan to court for going equipped to break into a car, get six witnesses who testify to this being the case, discover he's already got 22 convictions or findings of guilt (at 17 years of age), and what would you expect the sentence to be? Probably more than a fine of £1.60? Well, you'd be wrong, for that was the fine imposed by magistrates in May 1993, much to the disgust of local police officers and the witnesses who'd ended up sitting in court for hours.

In August 1989, a motorist from Coventry was on the run from the courts. Why? It could have had something to do with the 282 motoring charges he was facing. Magistrates added a further charge of failing to answer his bail and failing to turn up in court, as they issued a warrant for his arrest.

A Scottish Lord reached speeds of 110mph on the M4 motorway as he attempted to 'blow' the carburettor of his Maserati, a court in Newbury was told in August 1989. The Liberal peer was banned from driving for four weeks and was fined £120 with £15 costs. The Lord told the court he normally drove a somewhat more sedate Toyota.

You just couldn't put the brakes on a car-mad teenager, a court in Staines heard in September 1988. The young labourer, then aged 19, had never taken a driving lesson, let alone a driving test, and he didn't have insurance or vehicle test certificates either. But in nine

months he was stopped 20 times for motoring offences by police, sometimes twice on the same day, often just outside his door. Once, when a policeman served a summons on him, the teenager said, 'Just give me the stuff so I can go out again. You ain't gonna stop me.'

But when he went to court the magistrates certainly tried to. He was fined £3190 and banned from driving for 18 months after he faced 85 motoring offences. Magistrate Mrs Susan Walker said, 'This is the worst record we have ever heard of.' Afterwards, the angry teenager sped away from the court . . . on foot.

A Porsche is a pretty quick car, but John Hoyle's wasn't quite quick enough. He had been called to London's Capital Radio to fix a fault which had put the station temporarily off the air, but when he had finished and went back outside, his £30,000 pride and joy was being airlifted – by the police. It had been put in a sling and was being transported through the air to the back of a police truck prior to being carted off to the pound.

But John was having none of it. He leapt aboard the car – now 18 inches up in the air – and pleaded with police. They would not listen, however, and the car continued to rise into the air. John fired the engine, selected first gear – and drove off the airborne lift! A clamp was sent flying, the car crashed four feet to the ground, and John ended up at Wells Street magistrates court in July 1989 where he was fined £500 and ordered to pay £100 costs.

Businessman Andrew Stradling protested when he saw a shopper using his firm's private car park. But it was the parking offender, a 51-year-old woman, who really lost her cool when she returned to find her car blocked in: she burst into Mr Stradling's office and punched him in the mouth. She shrieked and swore while he was on the phone and threw office items about before socking it to him, Oxford Crown Court was told in January 1990.

MOTORING  *SHORTS*

Judge Peter Crawford told the woman, 'You are over 50 and ought to know better. Your conduct was like that of a three-year-old who lashes out after his wishes are frustrated.' He fined the woman £100 for common assault and ordered her to pay £400 in costs and compensation.

She'd left her car in the staff car park of a firm of chartered surveyors and when Mr Stradling tackled her about it she allegedly told him, 'I don't care. I'm going to do my shopping.' But when she returned 20 minutes later, said Anthony McGeorge, defending, her exit had been deliberately blocked. The woman claimed she had parked in the car park for only a few minutes to discover whether spaces were available in a busy local public car park.

A man charged with six motoring offences who would not take his flat cap off in court was locked up for two hours in August 1989, so he decided to get his own back when he reappeared before magistrates at Boston, Lincolnshire, wearing a white towel wrapped around his head like a turban. He refused to take it off, claiming he was an Indian Sikh. But the furious magistrates didn't see the joke, and the man was sent back to the cells for 10 minutes and fined £100 for contempt of court.

John Forster of Co. Durham told a doctor that he could take a blood sample from his big toe and nowhere else after he was charged with drink-driving in 1969. The doctor refused and Mr Forster was later acquitted of a charge of failing to provide a sample for a laboratory test.

The big-toe incident came after a police car had been forced into the side of the road by a car belonging to Mr Forster. He was arrested and taken to the local police station where they requested a blood sample. Dr Wallace Bexon said that when he asked Mr Forster where he would like the blood sample to be taken from, he replied,

MOTORING   SHORTS

'My big toe.' The doctor told him that this was not one of the accepted places, which were the ear, thumb, or arm.

The doctor told the court, 'He refused to allow me to take a sample from any of those places. I consider that his attitude constituted a deliberate refusal to provide a sample.' Dr Bexon added that he had told Mr Forster he had not wanted to take a sample from his big toe because of the risk of infection.

But Mr Richard Reed, defending, said, 'There would be a risk of infection wherever a sample was taken. It is not laid down anywhere in the Act that a person must give a blood sample from any particular place on his body.' The magistrates dismissed the charge. They also dismissed a charge of refusing to take a breath test after Mr Reed submitted that there was no evidence that Mr Forster had been driving the car.

Leading by Example is a worthy creed, and one which a good teacher surely must live by. So it was rather distressing to see that the headteacher of a school in Cumbria was charged with clocking his car mileometer in 1989. I shan't repeat the chap's name because I'm sure he suffered sufficient embarrassment when he had to go to court, where he was fined a whopping £700 for wiping thousands of miles off his car's clock. He also resigned from his Secondary School job.

You know the feeling. You're sitting in your car at traffic lights and someone draws up alongside in a hot Ford Fiesta. He's got his sunglasses on, music so loud it's making his little Ford tremble, and for the last three miles he's probably had his head out of the window – it's the only way he could have got that swept-back hair look. He wants to race.

Of course, being a sensible type, you ignore him, or perhaps you do rev your car up and get him going. Then when the lights change he squeals away from the lights – and you casually turn left down the side road. Most of us have been faced with this scenario, and experience

MOTORING  SHORTS

eventually dictates that the boy-racer should be allowed to go his own way and meet his Maker by himself. But, there are times . . .

And one such was in 1990 when an idiot motorist's duel with a taxi ended in the death of a pedestrian. The 30-year-old motorist was so intent on getting away from a taxi whose driver he had been arguing with in London's Piccadilly that he mounted the pavement and ploughed through a group of pedestrians, killing one, aged 65, and injuring two. The driver was charged with causing death by reckless driving as well as a further charge of reckless driving. Judges at the Old Bailey found the driver guilty – he'd pleaded not guilty – and jailed him for nine months.

Now take another look at the boy-racer alongside. Best to let him go on his way; it's not worth encouraging him, is it?

In 1993 when new laws came in to tighten up on non-payment of fines, there were few who didn't think it was a good idea – but all it did in many cases was make a complete ass of the law, so it wasn't too surprising when changes were made allowing magistrates to exercise common sense when fining, rather than having to follow often absurd guidelines issued by the Home Office.

Classic example of the fine-mania was in August 1993 when a motorist was fined £500 for failing to put a 50p pay-and-display parking ticket on his windscreen! But this was not the only example of the fining system gone mad, far from it. In May 1993 an Oxford motorist who twice popped out from his car in a pay-and-display car park without putting a ticket on his windscreen was fined a whopping £1200!

At the time, fines were linked to the ability to pay, and if you didn't fill in a means-test form, revealing details about salary and outgoings, then the court had no option but to assume that the offender had bags of money and therefore should be fined up to the hilt. The system couldn't last, but there was still time for a 20-year-old jobless man to be fined more than £1000 for dropping a

MOTORING *SHORTS*

crisp packet and for several magistrates throughout the country to resign in protest.

Police and the Crown Prosecution Service spent over £200,000 in a lengthy Crown Court trial brought over the non-payment of a mere 12 parking tickets, amounting to £144 in fines, in 1993. The man who found himself on trial faced 12 charges of perverting the course of justice (!) after police were unable to pinpoint which of his employees had incurred the parking tickets. He was found guilty of 11 charges and received a sentence of 180 hours community service. But he then appealed against the sentence.

The episode spanned more than three years, beginning with extensive police investigations into the parking tickets. In June 1990, the man was sent for trial by Warminster magistrates and the case at Swindon Crown Court in May 1991 lasted two and a half weeks and no less than 20 witnesses were called. The man's solicitor said, 'At a guess, the trial of this case alone has cost the taxpayer more than £100,000. The prosecution employed leading and junior counsel, we had junior counsel and my client is on legal aid. Guilty or not guilty, and obviously I cannot comment on the case itself, one surely has to question whether public funds are being used in the most appropriate way when cases such as this are pursued to such lengths.'

Indeed.

MOTORING  *SHORTS*

DON'T DRINK AND DRIVE

It's dangerous

As we all know, drink and driving do not mix, but sometimes drivers do not always abide by the rules. Here's a sobering look at what often happens after the night before.

There may well be a time and a place for chatting up girls – or men, for that matter – but it's definitely not the time or the place when you're driving a car at 85mph on a frosty main road after an evening of drinking. It's especially not recommended when you are still a learner driver.

All of these facts became clear to a man from Knottingley, West Yorkshire, when he appeared before Wakefield Crown Court in September 1989 charged with drink-driving, reckless driving and having no L-plates. The incident came to the attention of police when they saw the man and a pal on the A1 driving only eight feet behind another car which had a couple of girls inside. As his pal wound down a window and waved to them, the driver pulled alongside the girls' car and kept his crazy position for some distance before roaring off. Soon after, the 25-year-old show-off lost control of his car and skidded to a halt as he was chased by police. The court banned him from driving for two years.

The son of the head of the Del Monte food company, which makes soft drinks, was fined £2600 by a London Court in March 1993 for drink-driving. Italian Gianluca Sola, whose father Enrico was President of Del Monte Food International, was so drunk that he vomited when trapped inside a brand new Jaguar XJS by members of the public who stopped him from speeding away from the scene of a crash in London's West End.

MOTORING SHORTS

The 19-year-old student, who was more than twice over the limit, lived in Italy where there were no drink-driving laws, Marlborough Street Court was told. Magistrate Mr Quentin Campbell told him, 'You are clearly a rich and irresponsible young man.'

Not quite so rich after his court appearance, but it should have kept him sober for a while.

As an ambulance crew were inside a nightclub at Uckfield, East Sussex, in late 1993, battling to save the life of a man who had had a heart attack, a man drove the ambulance away from outside! A 21-year-old man was later charged with driving while disqualified and over the drink-drive limit. The things some clowns do! The heart attack victim, club member Robin Ward, fortunately later recovered in hospital.

An unemployed salesman from Leeds tried to drive off after downing 12 pints of lager – while his Ford Escort was still sporting a wheel clamp! He ripped a warning notice from his windscreen and set off down the road, his car not surprisingly making a loud grinding sound. As he crawled past a queue of late-night revellers waiting for taxis, he was stopped by two British Rail Transport police. While being breathalysed he admitted, 'I've had a couple.' In fact he was no less than three times over the limit and was jailed for six weeks in April 1993.

Two oafs from a top public school played dodgems with a stolen van, bouncing it off walls and buildings, a court heard in 1989. The two 17-year-olds, who were pupils at £2500-a-term Marlborough College in Wiltshire, then ploughed into a parked car causing £3500 worth of damage.

Marlborough magistrates heard that the driver was three times over the drink-drive limit. He was fined £550 with £25 costs and banned from driving for 18 months.

MOTORING  *SHORTS*

His passenger was fined £375 with £25 costs and banned for 15 months. The stupid driving spree happened as they celebrated the end of term. Both boys were expelled by the College, but look out, by now they may well have jobs in the City.

A party girl who was a little worse for wear moved her husband's car just 15 feet to let out another car from their drive – and was promptly arrested for drink driving. The young woman was being watched by nearby police officers, and not surprisingly they became somewhat suspicious when she jumped back in the passenger seat after reversing, and then promptly vomited all over the road. She escaped a driving ban at Blackpool court but was fined £500.

When Chris Day decided he shouldn't drive his Austin Allegro home because he'd had too much to drink, he little suspected that his old banger of a car would become, quite literally, a banger. Chris left the Allegro – nicknamed the Austin Aggro in Britain because of reliability problems – behind the Army Careers Office in Newcastle-upon-Tyne. Suspicious soldiers, always on the lookout for any IRA bomb-attacks, decided the car was a threat. They couldn't trace the car's registration, so local shops and homes were evacuated and when Chris walked to pick up his car the next morning it was gone. He reported it stolen, but in fact it had been blown up by the Army.

A Worcester man must have regretted the day he let his friend take the wheel of his car in March 1993. After a pub-crawl in the town, the friend got into a race – with the cops! He drove the car at 80mph, followed by pursuing police cars. The driver was eventually stopped and later taken to court where he was fined £924, plus £100 costs, while the car owner received a fine of £212 with £75 costs for letting him have the car!

MOTORING SHORTS

A farmer who drove to the police station to renew his shotgun licence ended up losing his driving licence. It happened because the policeman in the station smelt alcohol on the man's breath and asked him to take a test, which revealed that he was over the drink-drive limit. The farmer was later banned from driving for three years by magistrates in Wareham, Dorset, as well as being fined and having to pay costs amounting to £405.

'Fancy going to the police station to keep within the law,' he said, 'and then getting banned for breaking it.' Roger Curtis, defending, told the court that his client hadn't appreciated that the effects of the previous night's drinking would still be there. 'I told the police,' said the man, 'that I'd had a few the night before, and a hair-of-the-dog in the morning. But I didn't think I was over the limit, or I wouldn't have driven to the police station.' A sobering thought.

Three magistrates sitting at Brentford Court in 1969 had to decide which would dry out first – a man or a car. The man, a motor inspector of Southgate, London, was found sitting in his brand-new car in the River Thames at Isleworth, West London, by a policeman who judged him to be 'unsteady and quite drunk'. The man was accused at the court of being in charge of a car while unfit through drink, but he said he had parked his car near the river before going out drinking, and when he returned to the car the tide had come in and flooded it. 'I was so sick at seeing it in the river, I just sat in it,' he said.

The police officer who arrested him admitted, 'The car was so waterlogged it would have been impossible for him to have driven it.' The Chairman of the Magistrates, Mr VC Denton, said, 'We must decide whether the man would have sobered up before the car dried out,' the theory here being that he couldn't have driven the car when it was still submerged. Bad luck for the driver, though – the court decided that there was some doubt whether the car would have dried out before the driver would have sobered up. He was fined £20.

MOTORING  *SHORTS*

Here's a tale of misfortune, though it has to be said that the fellow did rather bring it upon himself. He was a failed businessman from Warrington and he was jailed in March 1990, just 10 days before his wedding, after magistrates heard that he drove when more than three times over the drink-drive limit, and while banned for a previous offence. The court heard that he also had a previous broken marriage, had put his money into a failed business, and had a big drink problem. Trouble was, when he went to court to answer the charges, the 45-year-old drove himself there, and even drove home afterwards – while he was still banned from driving.

'This was one of the most serious offences of deliberate disqualified driving the court has seen,' spluttered the magistrate, but the man's solicitor said he had been anxious to be on time at court, panicked, and driven himself when a lift failed to arrive. The businessman was jailed for six months and banned from driving for a further three years. More misfortune there.

Watch what your friends are giving you to drink – that's the message following a court case in Scotland where it was revealed that a man who thought he had drunk four pints of low-alcohol lager – in an hour, it must be said, what a thirst! – had in fact been handed four pints of normal strength booze, and was later stopped by police for drink-driving.

The man told the court how he had been drinking with his friend and he had definitely asked for low-strength LA lager. But, as his pal told the court, he had misunderstood and actually bought ordinary-strength lager, without telling the car driver. Despite this tale of woe, the drink-driver was still banned at Haddington Sheriff Court in Scotland for three years and fined £125.

MOTORING  *SHORTS*

FAMILY AFFAIRS

Sometimes you don't know who your friends are

Family and friends may seem the closest people you have, the ones you can rely on, the ones who value you the most, but when it comes to motoring . . .

Ian Davis from Broadstairs, Kent, didn't do his two-year-old son Oliver any favours when he failed to stop his car in time while parking it. The car lurched forwards and knocked down the garage wall. Unfortunately, young Oliver was behind the wall playing in the back garden. But fortunately he escaped with only minor cuts and bruises.

'It was a terrible feeling,' said Ian. Hmm, I suspect it wasn't a particularly good feeling for Oliver either.

A female learner-driver's joke on her boyfriend turned into a crashing disaster as she damaged two cars and demolished a garden wall, causing £2000 worth of damage. As well as losing her boyfriend, she also lost her driving licence.

The oil-rig worker left the girl sitting in his BMW with the engine running when he popped into a Chinese takeaway in Middlesbrough in March 1989. She decided it would be fun to hide the car – the man's pride and joy, naturally – around the corner, but she pressed the accelerator instead of the brake. The BMW smashed into the back of a parked Ford Fiesta and then careered into a garden wall. The angry householder dialed 999 and the poor girl was arrested, breathalysed, and charged.

Magistrates fined her £100 and banned her from driving for 18 months after she admitted driving carelessly, and with excess alcohol, having no insurance or L-plates and being unsupervised. She also faced damages claims for £700 for a new wall and £680 for repairs to the car she hit.

MOTORING SHORTS

The BMW driver paid off the £500 bill for damage to his own car but called off the romance. His now ex-girlfriend said, 'It was only meant as a leg-pull but it turned into a disaster. I drove 100 yards, but it was the costliest trip of my life. To make matters worse, I picked the worst possible spot to have an accident – right opposite a police station.'

It certainly was one of those days.

Architect Rob Mitchell faced jail – and all because of his wife's war with prowling traffic wardens. Mel Mitchell had said she would rather go behind bars than pay penalties for parking on double yellow lines outside the couple's home. But Rob owned the car, and he was legally responsible for any tickets stuck on its windscreen. By December 1989 he owed more than £1400, and he said, 'It's all very well for her to stick to her principles – she isn't the one who will have to go to prison.'

But unrepentant Mel said, 'I get the tickets – and he has to pay them. Anyway, he earns more than me.'

The problem stemmed from council regulations which meant that the couple could not buy a parking permit for St Mark's Road, Widcombe, Bath, and there was nowhere else for Mel to leave her car during the day. 'It's ridiculous,' she said. 'If we lived 50 yards down the road we would be able to buy a permit. A parking ticket is a daily occurrence. The traffic wardens are like part of the family.'

Rob told Bath magistrates that he would pay off some of the fines rather than spend that Christmas in prison!

A caring husband from the Isle of Wight who tried to stop his wife from drink-driving soon realised he had made a serious error of judgement. When she refused to hand over the car keys, he flung himself on the bonnet of their car. Trouble was, she ignored him and continued to drive the car through the streets, with him clinging on desperately!

The boys in blue duly spotted him behaving rather strangely on the outside of the car, and gave chase. When they finally caught up and stopped the vehicle, his wife was breathalysed, leading swiftly to her arrest for drink driving, a charge to which she later pleaded guilty in court. A charge of using a vehicle with a passenger in a dangerous position was dropped. Magistrates banned her for a year and fined her £200 with £10 costs.

Jim Gatwood, a used-car dealer in Christchurch, New Zealand, made what can best be described as an error of judgement when he sold his mother, Joan, a second-hand Mini. Joan later found that her loving son had charged her four times the real value of the car!

'I love my mother, but business is business,' said hard-hearted Jim. Joan didn't quite see it that way. She decided to picket the used-car lot, telling other potential buyers about the dastardly deed, until Jim refunded the money.

Keeping a picture of a baby on the dashboard of your car makes you drive more safely, according to Austrian motoring organisation Arboc. 'It arouses protective instincts in the driver,' said a spokesman. Presumably though, you shouldn't be looking at Junior while you are driving.

Nine-year-old David Dillman shouldn't have been at the wheel of a car, and he certainly shouldn't have been driving his mother's van down the San Diego Freeway in Los Angeles, one of the world's busiest roads. But it's a good thing he did, because it saved her life.

When Mrs Mary Dillman blacked out at the wheel in the fast lane of the five-lane highway, David quickly unbuckled his seatbelt, clambered onto his mother's lap and then had to swerve the van through heavy traffic as he battled to keep control. 'I just kept thinking my mom was going to die, and I just had to get the van over to the

MOTORING  SHORTS

right-hand side of the road,' said David in March 1994.

Luckily, David's mother had shown him how to drive when he was seven. But even that almost didn't help as at one point he managed to put his foot onto a pedal – the accelerator pedal! 'But then I knew it was the other pedal I wanted,' said the intelligent young man, who eventually managed to stop the van on the hard-shoulder. Mrs Dillman was later released from hospital, with doctors still unsure why she blacked out.

Alain Prost, the recently retired Formula One racing champion, is not a man known for his accidents; rather he is known for safely winning races. But his wife Anne-Marie did have something of a shunt, as it were, back in 1990. She was seven months pregnant at the time, and her car was hit by a train! Driving across a level crossing in Switzerland, she was shunted down the track for several yards. She and her son Nicholas thankfully escaped unhurt, as did the unborn child.

MOTORING  *SHORTS*

MOVING IN MYSTERIOUS WAYS

Not so much God himself, as some of his servants down here on the roads.

A religious fanatic in Geneva, Switzerland, was saved from a traffic conviction after he called God as his witness. When the police stopped the driver, he said, 'God was driving my car. I was on my way to buy condoms to stop an AIDS epidemic.'

The driver refused to give evidence until the Swiss court allowed him to call God as a witness. It appears the court was full of non-believers; the case was dropped.

An elderly Sister of Mercy dedicated to helping the poor, needy and handicapped admitted to failing to stop after a hit-and-run road accident which left a pedestrian unable to walk without the aid of sticks. The 77-year-old Sister, of the Order of the Sisters of Charity of St Vincent de Paul at St Catherine's Convent, Lanark, Scotland, also admitted failing to report the accident to police. When she appeared in court in April 1990 she was given a six-month driving ban and ordered to retake the test if she wanted to drive again.

Here's a plan which may well have had the Good Lord grumbling. It was an idea to build a car park on top of a new purpose-built church, and to call various floors of the multi-storey car park Heaven, Hell and Purgatory.

The plan came about in January 1990 in Chicago, mainly as a result of church economics and the problems of finding somewhere to park in Chicago. The city's oldest Roman Catholic Church, St Mary's, added 'and Self Park' to its name, and according to plans submitted to Chicago City Council, the original church building was to be demolished and replaced with a modern

MOTORING SHORTS

12-storey structure – 10 floors of parking on top of a two-tier church, in place of the somewhat more traditional church spire! There was to be parking for 850 vehicles, and an ersatz bell-tower, topped with a stainless steel cross, would contain the lift to the parking floors. Down below were pews for 450 worshippers – and free parking on Sundays and selected Holy days for churchgoers.

'The building will look like a church,' insisted the project's architect, Mr William McBride. 'We want the parking garage to look secondary.' The congregation was aware that it would have trouble raising the $3–4 million needed to rebuild the old church, and under the proposal for the new building, parishioners were to get a new church worth $5.5million for virtually nothing – and they'd receive a 2.5 percent share of the parking receipts. From the point of view of the developers – Standard Parking – it was also a good deal because they got lucrative city centre parking.

The Rev. Thomas Dove, the parish priest, revealed that the collaboration with Standard Parking had led to a debate on whether the parking floors should have individual names. Two of the company's other parking complexes in Chicago played on themes. One in the theatre district had floors named after musicals, with tunes from the shows playing on each level; another had floors with US city names and theme tunes to identify them. 'We gave up on finding 10 saints in Chicago,' said Reverend Dove. His other suggestion was that the parking levels be named Heaven, Purgatory and Hell. In the end, Standard Parking decided that religious references were 'inappropriate'.

A good Samaritan priest gave a man a lift while three times over the drink-driving limit, and it ended in a trip to court. The helpful Father had been drinking wine in a social club near Chiswick, West London, when a lost man came in. 'It appealed to his Good Samaritan nature,' said James Lewis, defending, 'and he used his car to give this distressed young man a lift home.'

Unfortunately, he was soon in distress himself as

MOTORING *SHORTS*

police spotted his Vauxhall Astra swerving along Streatham High Road at 4am. The priest admitted drink driving, was banned from driving for a year and fined £300 by magistrates. 'He has not drunk any alcohol since the incident,' said Mr Lewis at the court case in April 1990.

It's said that God moves in mysterious ways; and it's certainly true of some of his mortal servants: Canon John O'Flaherty caused an almighty mess when he drove up the M4 the wrong way after getting lost. He and seven others were slightly hurt in several resulting collisions on 2 August 1990 at Bridgend, Wales.

LEARNING TO DRIVE

Madly-beating hearts, sweaty palms, sudden rushes of sheer panic or fear – and that's just the examiners . . .

In 1990, driving examiners made a call for Scots and Sikhs who take their driving tests in national dress to be banned from wearing daggers. In their profession they had become philosophical about violence at the hands of failed learner-drivers, but clearly they preferred candidates not to come to their driving tests with a dagger strapped to their sides. The call for a ban followed a driving test in which a Sikh in national costume drew his dagger to 'emphasise his disappointment' over failing.

The Department of Transport expressed its concern: 'We have sent letters to community relations officers and race relations councils seeking their views on the possibility of imposing a ban on these daggers,' said a spokesman, who added, 'It is not unusual for candidates who have failed their test to become abusive.'

But experts didn't believe that really committed nationalists would give up their weapons so easily. Mr Danny MacLean, then Secretary of the Gaelic Association, said, 'There are Gaels who are fanatical enough to insist on wearing full dress, including the scian.' And Mr Balraj Singh Dhesi, chairman of the Community Relations Council in Leamington Spa, said devout Sikhs would not lay down their daggers 'under any circumstances'.

Of course, it's not so much the laying down that the examiners are worried about . . .

A driving test centre in a quiet market town was forced into an emergency stop – because light traffic made it too easy to pass the driving test! Learner drivers cruised through examination routes in Machynlleth, North

Wales, unaware of the perils and pitfalls of clutch control and traffic congestion.

Roads and Traffic Minister Robert Atkins said in April 1990, 'The test routes in the town are not a realistic test of a candidate's ability to drive.' Apparently, learner drivers had been flocking to the quiet Welsh town in an effort to pass their tests with the minimum of effort. But the centre closed in June 1990, which meant that local learner drivers had to go to Aberystwyth, or Newtown and Bala.

However, one man who hadn't found the town's test too easy was local carpenter Richard Jones, who said, 'I've failed my test six times over the past 10 years. I certainly don't think the roads here are too easy.'

Believe it or not, Australian women who dress in scanty clothes for their driving test are deliberately failed, according to a Sydney driving instructor. And quite right too; must maintain the proper standards. But there's a catch.

Apparently, according to the female instructor, they are failed by some of the male examiners so they can get another look when the girls re-take their test! It could only happen in Australia.

Twenty-stone Daniel Cowdrey couldn't wait to get started when back in August 1989 his mum bought him a course of driving lessons for his 17th birthday – but when he tried to get behind the wheel, massive Daniel wheezed and squeezed but just could not get his giant frame into any of the cars the driving school offered him. He's 6ft 10in tall and takes size 16 shoes.

'I either have my knees jammed up underneath my chin or I have to stick my head out of the sunroof – which is not only dangerous but looks daft,' said Daniel. He began his lessons by trying to squeeze into a Mini, of all things, at the British School of Motoring and ended up trying a Ford Sierra. But even that was too cramped.

Daniel's mum Irene, of Basingstoke, Hampshire, said, 'Just getting into a car is a big problem for Dan. And when he tries to drive he is so scrunched up that he can't push the pedals and look through the windscreen at the same time.

'He's such a big lad that we already have to get his clothes and shoes specially made. Now it looks like we'll have to do the same with a car.'

Debra Brough will never forget her first driving lesson. Seconds after she drove off, instructor Ken Shaw had a heart attack. But as he fought for breath, Debra did the only thing possible, she put her foot down. She drove two miles through busy traffic in Blackpool to reach the Victoria hospital – and saved her instructor's life. 'It was only when I got there I realised what I'd done. Then I started shaking like a leaf,' said Debra in August 1989.

A driving instructor most definitely forgot the Highway Code when he fled from two policemen who he thought were thugs.

The instructor from Chelmsford, Essex, appeared in court in July 1989 charged with reckless driving. He went through two sets of red traffic lights, drove at 60mph in a 30mph zone, and 100mph on a 70mph motorway. Then he tried to overtake as oncoming cars came towards him. He was chased for 16 miles by the officers in the unmarked police Ford Granada.

The driving instructor claimed, not without some justification it would seem, that he thought he was being pursued by two thugs. Asked in court whether he taught pupils to stop at junctions, the instructor protested, 'They are not pursued by a couple of maniacs.'

When Rosie Milvain passed her driving test she could not have been happier. At 84 years of age, she was the oldest person in Britain ever to pass the test. But disaster struck.

MOTORING  *SHORTS*

Just the next day her brand new Daihatsu Fourtrak, with her two dogs inside, was stolen from a supermarket car park. Local radio stations asked listeners to look out for the heartless thieves, the police scoured the area, an extensive search followed, and local people were aghast at the horrible crime. How could anyone do this to the poor old woman and her dogs?

But hours later the police discovered the culprit. It was Rosie herself. Britain's oldest new driver had left the Daihatsu in another nearby car park and forgotten where it was parked! Rosie, who had driven for an accident-free 67 years, had been ordered to take the test after being fined by a court for careless driving.

Learner driver Joan Eadington passed her driving test in January 1994. Nothing too surprising in that, you might think. Well you'd be wrong, because Joan had been learning to drive over a 20-year period, and had taken, and failed, no less than 20 driving tests.

'At times I was absolutely desperate,' said 67-year old Joan, who started to learn to drive after winning a Vauxhall Victor in a competition. Vauxhall ceased production of the Victor back in the mid-1970s.

Just two hours after buying his car, learner driver Glenn Aldred drove it straight off a quayside at Portsmouth Docks, flipping it over and landing it against the side of a pilot boat anchored to a pontoon down below.

'I was parking ready for my driving lesson,' said Glenn, 'when the car just headed for the edge of the quay and went over. I was going to take my test in that car in six weeks.'

'Fraid not, Glenn. But perhaps they might consider letting you take the submarine driver's course . . .

Driving instructor Jan Groot from Amsterdam slips a whoopee cushion under nervous pupils to break the ice on the day they take their driving test.

MOTORING SHORTS

Learner driver José Quintauilla from Leon, Spain, not only failed his test for the second time but was also presented with a bill for 16,000 pesetas. During the first test he had hit a bus, three parked cars, and a supermarket. Now in his second attempt he hit a traffic light and a lamppost, breaking his examiner's leg in the process.

Apparently, word spreads like wildfire when José is taking a driving test, and many locals stay indoors – though those who hide in the supermarket are unlikely to be safe from his exploits.

MOTORING  *SHORTS*

POLICE SQUAD

They're on patrol, but don't always get their man

The police are not always seen in a favourable light by the motorist, usually because the boys in blue clamp down on speeding and reckless driving and the like. Sometimes, though, the police make mistakes too . . .

Bungling coppers once towed away a two-year-old boy when they removed his mother's parked car. They didn't discover sleeping David Boahen until 90 minutes later when distraught mum Lorna raised the alarm. She had left her car on the pavement in Kennington, London, while she visited a friend.

'David was asleep, all the doors were locked and I didn't think I'd be away long,' she said, 'but when I returned, the car was gone. Police eventually rang the pound and found David, who was really frightened.' Mrs Boahen decided to appeal against the £69 fine.

Even Chief Constables cannot escape the long arm of the law, at least Keith Povey, head of the Leicestershire force, could not – he was caught by his own officers as he sped through a 50mph section of motorway roadworks at 80mph. In court, Povey was banned from driving for a week and fined £350.

Police had to chase one of their own when a jilted constable raced along public roads at speeds of up to 105mph in an effort to stop his girlfriend leaving him. The late-night chase in 1989 ended with the officer's car wrecked and two police vehicles damaged.

The besotted 38-year old officer, who had left his wife and children to set up home with the woman, was on duty when she told him over the phone that it was all over between them. He roared off in a bid to save the relationship, ignoring red traffic lights and making other vehicles swerve before crashing into a police roadblock manned by his colleagues.

Leeds magistrates heard how the devoted father and respected officer was later reconciled with his family, but had not felt able to drive or go back to work. He was banned from driving for two years and fined £100 after admitting reckless driving. The police suspended him from duty and made him face a disciplinary inquiry.

A diminutive British Special Constable flagged down drivers for imaginary motoring offences and then charged them a fiver to be 'let off'. The blond five-footer was turned down by the full-time force for being too short, but he was accepted as a part-time Special and, dressed in his Special's uniform, he pretended he was a full-time traffic cop.

The man from Burntwood, Staffordshire, found himself jailed for a year in 1989 after he admitted four charges of blackmail. David Crigman, prosecuting, told the court how the Special Constable asked one woman he stopped, for supposedly speeding at 70mph, for £5. Another motorist was apparently told, 'You were driving in such a way that you will get a substantial fine. If you buy me a drink I might let you off.'

The man was a Special Constable for three months and when he was eventually collared he at first tried to blame friends who were with him. His lawyer, John Mason, said, 'He can find no sensible explanation as to why he did it – it was not through lack of finance or out of a desire to hurt people.'

Policemen and women in Britain are grateful for the new longer 22-inch truncheons, which help them keep the violent criminal at arm's length. But there's one problem. They have to remember to remove them from their pockets before they get in their patrol cars, otherwise they cannot bend their knees to drive.

There have been several accidents thanks to the extra-long truncheons interfering with police drivers' braking at crucial moments – like when they want to stop! – and at least one constable in London's Metropolitan Police had to receive hospital treatment after he jumped energetically into his patrol car and was injured in a most delicate area by the big truncheon.

'He will make a full recovery,' said a police spokesman, 'but he has asked not to be named in case it affects his chances of getting married and eventually having children.'

MOTORING  *SHORTS*

TAKEN FOR A RIDE

Here come the car thieves

Theft from cars and the theft of cars is unfortunately very common these days, but sometimes the thieves don't always come out on top . . .

Two Irish car thieves put themselves in the frame when they proudly took photos of each other standing beside their stolen autos! The cousins were caught after they handed in their snaps for developing and were spotted holding their break-in kits by eagle-eyed staff at the photo-processing company.

The police soon tracked the Irishmen down and recovered the cars. They were later sentenced by Bournemouth magistrates to do community service and were also banned from driving. Smile, please!

The moral of this story may well be that you never, just never, leave your car. It was an incident that happened back in 1969 when a Mr DA Stoddard of Atlanta, Georgia, discovered that his car's battery had been stolen and the petrol tank drained overnight. So, not unnaturally, he took himself off to the local garage to buy another battery and more petrol. But when he returned he found the two front wheels missing.

By now you'll be getting the gist of this disaster. Yes, incredibly he went away to buy two spare wheels and returned to discover, guess what, the whole car had vanished. Reporting his loss to the police, he learned that a policeman, seeing the partly stripped car, had assumed it had been abandoned and made arrangements for it to be towed away. Well, at least it was safe – what was left of it.

Labour MP Ann Clwyd is not too attached to her car, or

MOTORING *SHORTS*

at least she wasn't in April 1989 when she reported it stolen to the police, telling them it had been taken from the House of Commons car park. In fact, the car had not been stolen at all: Ms Clwyd had simply forgotten that she had left it locked up at Paddington railway station some two weeks earlier!

Mind you, when the story was reported in the *Daily Star* the reporter admitted that a similar thing had happened to him only a few days earlier. As he was battling his way through the rush-hour crowds at London's Waterloo station in an effort to board a train home, he suddenly remembered that he'd come to work that morning in his car, and it was still safely tucked up at the *Daily Star* office car park. He said his momentary lapses of memory usually occurred after 'a heavy night on the booze, but they rarely last more than a few minutes'.

And these people are out driving around on our roads? Well, only when they don't forget where the car is parked.

Customers frequenting Hollywood's Top Hat restaurant have been making a bit of a boob recently. Many of them have been handing their car keys over to the valet. Nothing strange in that, you might think, but it's always wise to read the sign outside first. 'Check the face of the valet who parks your car,' it warns, 'as we do not have valet parking here!'

Sound advice, and just as valid across the pond as this story shows. A car salesman from Manchester, England wanted to impress his new girlfriend, so he whisked her off to an expensive restaurant in his flashy high-performance Lotus Esprit Turbo. He'd only had the car for a mere 48 hours as he jumped out, left the keys with the valet and took his girl inside.

As they sat down for the meal, everything was perfect: the food, the wine, the ambience. It looked like a night to remember. And it was, but not for any good reason. As they sipped their wine, there was a phone call for the

MOTORING *SHORTS*

man – from the 'valet' who'd just taken his Lotus. Yes, you've guessed it, this restaurant didn't have valet parking!

'It was very embarrassing, especially as it was my first date with this girl,' said the victim, who was too embarrassed to be named. 'The thief told me I couldn't have the car back because I'd handed over the keys.' A spokesman for the restaurant said, 'All our attendants carry clear IDs, wear distinctive uniforms, and issue receipts. They don't work on Sundays, when this customer came into dine.'

If you are one of the increasing number of unlucky car owners who have had their car broken into or stolen, spare a thought for Brian Rankin of Wakefield, West Yorkshire, who was the target of no less than 14 break-ins in only two years in the early 1990s.

During this spree, thieves took jewelry, stereo systems, birthday presents and clothing. The thefts happened all over the country and the value of the stolen property amounted to more than £2000 by January 1991. Mr Rankin couldn't believe it. He began taking all kinds of precautions.

'I have had cars with sophisticated alarms and cars without alarms,' he said, 'but it makes no difference. At one time I started leaving my car unlocked simply to avoid damage to the bodywork, but the insurance company wasn't too happy.'

It's at times like this that public transport begins to look pretty exciting.

THE WRITE STUFF

A journalist tells tales

The life of a motoring journalist can be fun. There are new cars to drive every week, and some of them are the sort of machines that only the very richest drivers can ever afford. And every time a car manufacturer launches a new car, the motoring scribes get invited on all-expenses-paid trips to some of the world's most exotic locations where they are wined and dined in the world's most expensive hotels.

This may seem rather obscene and sometimes it even seems that way to the journalists as well, but there's no doubt that it does make some sense to take the scribblers away somewhere warm and sunny which allows them to sample a new car properly and then hopefully, from the car-makers' point of view at least, come home and write a glowing report.

Now, lest you think that journalists' opinions are, heaven forbid, shaped by the trip they have just been on rather than the car itself, remember that the journalists are going on trips like this all the time – the joke in the industry is, oh yes, it's Tuesday, we must be in Biarritz – so it doesn't really affect their judgement when they're writing about the cars (well, it's never affected mine!).

The point is, you get a plane load of journalists, send them somewhere very pleasant . . . and all sorts of things happen. Some of the more amusing episodes, I thought I would share with you.

Obviously, the car makers do try to outdo each other to some extent, picking more and more desirable locations (on one famous occasion in the mid-1980s, Rover claimed that they were only taking journalists to sunny Nice, in the South of France, because there was snow on the ground in England which would have hindered proper road testing of the car; quite how Rover knew it would be snowy in England when they planned the trip a month or so earlier was something of a mystery).

The other element which is added into these trips is what the car-makers usually refer to as 'a small gift to remind you of your trip to blah-blah'. Perhaps not surprisingly then, the journalists refer to this present as 'The Bribe', and it has ranged from a pair of initialled Waterman pens (relatively inexpensive) to a complete bone china dinner-service, though the latter was somewhat unfortunately stamped with the car company's name on the bottom.

I know this because one chap told me about it after he'd given the dinner service to his mother-in-law for Christmas. 'Oh, how thoughtful, dear,' she said, before turning one of the plates over to look at the bottom to read: ' "From Audi. Vorsprung durch technik". Hmm, never heard of them, but what good quality china this is.' The remainder of the Christmas holiday consisted of the journalist's wife giving him severe how-could-you looks.

Today, The Bribe is neither as lavish nor as common as it was. Indeed, on some trips there is no Bribe to be seen at all. Disaster! But back in the mid-eighties there was always a Bribe. Except for one memorable occasion when the powers-that-be decided that there would be no present, the trip was quite sufficient. But the journalists didn't know this. And what happened next is absolutely true . . .

It was a trip to France in 1985 to test a new French car, a sporty version of a more mundane family hatchback. On the trip was the then-usual assortment of journalists, comprising of old boys who had flown with the RAF in 'Spits' during World War II, plus a handful of younger journalists like myself. There was also a gaggle of public relations people, both from the French parent company and from their British operation.

So, we drive the cars, we stay in one of the world's finest hotels where we have some glorious food, wonderful service and views of the ocean which you normally only find in a brochure. Then over dinner, one of the old buffers pipes up, 'Can't seem to find me present. Nothing in the room.' They all start grumbling about it, until one of the younger hacks – not myself, I hasten to add – says, 'What? Haven't you seen it?'

'What?' say the older ones in unison.

'Yes, it's in the corner of the room. Well, mine is.'

They all look puzzled for a moment and then one of them remembers it and they are all happy again.

The following morning a procession of the older journalists joined us on the bus to the airport clutching their 'Bribes'. Each one of them had trooped past the reception desk clutching a tall rubber plant that they had just stolen from their hotel rooms. The hotel manager was so taken aback by the bare-faced thieving – or perhaps he thought it was just some quaint English custom and besides, the car company has paid a fortune for the rooms – that he just shrugged: what's a few rubber tree plants?

We younger chaps, already on the bus (minus the hotel's rubber plants of course), were finding it increasingly hard to keep a straight face as these older journalists clambered on board with their tall plants. 'Ha,' said one, 'forgot yours, did you?' and they all started laughing at our inexperience. We laughed with them, by now quite beside ourselves at how absurd this lot were.

Of course, once we got to Heathrow they were all stopped as they strolled through customs like a mobile rain forest, and their plants were denied entry to the UK, much to their annoyance: 'I'll have you know, this is a present, my man,' one of them was heard to bluster.

Another of my particular favourites is the story of the launch of the Fiat Mirafiori, back in the late 1970s one of the first cars with a five-speed manual gearbox. The launch was held in and around Rome, and after the

MOTORING   SHORTS

driving on the second day, in the early evening, one of the oldest car journalists – his name was Geraint and he was often referred to as The Geriatric – who mistakenly believed his eyesight and judgement to be as good as the day he was in Rome with the liberating British army during the last war, was driving a carload of other journalists to a restaurant. Somehow he'd managed to get behind the wheel, which was a miracle in itself, for two reasons.

First, he had a walking stick which he always placed by his side in the car and which he had been known to mistake for the gear-lever on occasion – 'Bloody thing's stuck, can't get it into third' – and second, other journalists always tried really hard to be in any other car than the one he was driving, such was his lack of skill and judgement at the wheel. Anyway, on this occasion, there he was threading the sporty Mirafiori through the back streets of Rome at hectic speed. 'Don't worry,' he told the terrified passengers, 'I know me way around, drove a tank through here in '44.'

Trouble was, back in 1944 some of these streets that he was now charging up had not been one-way, or if they had, I suspect it didn't particularly matter if you were driving a British battle tank. On he went, often driving against the traffic, going up one-way streets the wrong way, much to the consternation of the locals – 'Chaps are always hooting their horns in Rome, wonderful' – and with his passengers by now shocked into silence, praying hard that they would soon reach the restaurant.

'Twas not to be, as The Geriatric then made the mother of all mistakes. He really had the Mirafiori flying and he was in fourth gear. He was jabbering away, swerving in and out of traffic, when he wrenched the gear-lever out of fourth . . . and slapped it straight into first. The Mirafiori stood on its nose, various cogs and parts of the transmission system broke away and went skittering up the road, The Geriatric was half flung out of the open side window, the passengers in the back joined those in the front and the car eventually skidded to a halt, firmly embedded in the front of a fruit shop.

Fortunately, and somewhat against the odds, there

MOTORING  *SHORTS*

were no serious injuries, though The Geriatric was not invited on a Fiat trip again.

I once had to go on a Toyota trip to Scotland – well not once actually, because at the time almost all Toyota trips were to Scotland, their PR man back then being Scottish and liking nothing better than to show his homeland off. Unfortunately, he also had a penchant for getting us on all sorts of bizarre aircraft.

In this particular case it was a twin-engined Otter (one other time it was a Dakota flight to Spain, and halfway there one of the engines burst into flames) flying from Biggin Hill airfield to Dundee. Now, I hate flying anyway, but the thought of being cooped up in a small turbo-prop plane for a couple of hours didn't thrill me to bits. Looking nervously out of the waiting-room window, I asked one of the other journalists if he knew what a twin-engined Otter looked like.

The chap I asked was, inevitably, an ex-RAF pilot, so he must know about these things, I thought. He pointed out of the window of the waiting room – at Biggin Hill then it was just like a large garden shed – and said, 'See, there's a twin-engined Otter taking off.' I watched the small plane as it took to the air – and then plummeted to the ground. 'And that,' said my colleague grimly, 'is a twin-engined Otter crashing.' This was not good news, because two people on board the plane died. Still, this was not our Otter.

Next thing we knew, there was a pilot named Jim, who looked about 16, telling us that, 'This is Mandy, your air hostess. Now in the event of an emergency, don't listen to Mandy, you listen to me' – a little smile here, I remember it like it was yesterday – 'because we don't want any confusion, do we?'

We got on the plane and for some reason everybody sat at the back. 'Don't you all want to spread yourselves out a bit?' says Jim. 'It makes it a little easier to take-off.' We all split up, spreading ourselves across the seats, anything to spread the weight. We took off and started trundling up through the clouds and round came Mandy to see that we were all OK. She took one look at me and my white face and asked if I was alright. I told her I didn't much like flying and she had an idea: 'Why don't you go sit up front with Jim, it always makes people feel better.'

'Won't it be a bit crowded up there with Jim and the co-pilot?'

'Oh, silly,' she laughed lightly, 'we can't afford a co-pilot.'

I swept aside the curtain which separated us from the cockpit and went through, easing myself carefully into the co-pilot's seat, careful not to touch anything. Jim handed me some headphones and I put them on. We were still climbing up through the clouds and then all of a sudden, right there in front of us, was a glider.

Jim cursed and stamped on this big pedal on the floor, sending the plane into an immediate nose-dive. I heard the rush of many feet from behind and all of a sudden four or five journalists, complete with drinks, were wedged in the doorway.

'Steady on, old chap,' said one of them, as Jim levelled out and started to climb quickly – and they all disappeared like magic, hurtling backwards into their seats again.

'Jesus, that was close,' sighed Jim.

'Didn't he show up on the radar?' I asked.

Jim gave me a broad smile, then leaned forward and flipped a switch up.

'He will now.'

The rest of the trip was uneventful until we approached Dundee airport and got a message from the RAF that they were carrying out low-flying missions and we should climb to get clear of them. We did so and then Jim started looking down out of the window. Worried about the RAF planes? No, he said quietly. But he kept looking and then he said, 'You been to Scotland before?'

'Not really,' I replied, 'only once or twice when I was a kid.' He said nothing, just kept looking down.

'Never been to Dundee, then?'

'No, of course not.' The first faint stirrings of worry in my mind now.

'Ever seen a picture of it?' That broad smile again. We were lost!

Eventually I spot a large piece of tarmac which turns out to be Dundee Airport. We are meeting Brian, the Scottish PR man at the airport, and I haven't met him before. (I should point out at this stage that my first name is actually David, not Tony, and this is the name that Toyota have put on their invitation.) I get off the Otter and Brian greets me.

'Welcome to Dundee, David.'

'Tony,' I say.

'No,' he smiles at me. 'My name's Brian.'

This trip just didn't get any better.

There's one particular journalist who is invited on numerous press trips but hardly ever manages to turn up. It's not that he doesn't want to go, it's simply that he seems unable to a) get out of bed on time, b) arrive at the right airport, or c) . . . well, let me explain about one particular trip.

It was a Ford trip to Ireland which was just for the day. Catch the plane early in the morning from Heathrow, fly to Dublin Airport where the cars would be ready and waiting (I seem to recall it was a new version of the Ford Escort), drive along a set route to a beautiful hotel out in the country, have a first-class leisurely lunch, drive again in the afternoon, then fly back to London in the early evening. A nice little trip – if you actually got on it. Our man – let's call him Arnold, though I hasten to add this is not his name – gets out of bed late. He stumbles into his clothes, gets in his car, and drives to the airport. Gatwick Airport. Unfortunately, the Ford flight has already left – from Heathrow.

However, undeterred, our man Arnold sees there's a flight about to leave for Dublin. Hah, he says to himself, I'll only be a little bit late. And he has an itinerary – these are always sent out prior to the trip – so he knows the address of the hotel, though he doesn't have the route to it – these are always given when you get into the cars. So he gets on the flight. Success. He lands in Dublin. Of course, there's no-one from Ford anywhere to be seen, they've already set off with the other journalists.

But Arnold has the bit between his teeth now. He jumps in a taxi and asks to be taken to the hotel. But it's a long journey – always on press trips the driving routes are long, usually around 100 miles in the morning, another 100 in the afternoon – so this taxi trip will cost a fortune. Not only that, by the time he eventually gets to the hotel, all the Ford people and all the journalists have already had lunch and set off on their afternoon trip.

Arnold has a sandwich, and then gets a taxi back to Dublin. He gets on an evening flight back to Heathrow – which is a shame, because his car is at Gatwick. He's spent all day travelling, but, in a manner of speaking, he never actually arrived.

It was in the mid-1980s that car companies started to spread their press-trip wings a little, and instead of the usual trips to the South of France, Germany, Monte Carlo or Spain, suddenly there were even more exotic climes. One of the first such venues was Turkey, which at that time was not the popular holiday destination that it is now. So there was some trepidation on the part of the journalists who were invited to Turkey, as they worried about various things foreign which they might not have experienced before.

The worst of these fears was food poisoning, a fear underlined by the old hands, some of whom had got the bug during their wartime service (mostly at Aldershot, but that's another story). So, off we all went to Turkey and everyone was oh-so careful about what they ate and drank – no sheep's eyes, no mutton. The trip, which was for a whole week (yes, we did drive the cars extensively, though I'm blowed if I can remember what they were, or even which company took us – sorry!) went really smoothly, no-one came down with anything and there was an almost audible sigh of relief when we sat down for a meal on the British Airways flight back home.

Right first time: just about everyone got food poisoning from the in-flight chicken.

MOTORING SHORTS

You may remember Steve McQueen starring in the cult movie Bullitt, where a large part of the film is taken up with McQueen involved in a long and exciting car-chase – the ultimate car chase, you might say.

Well, there's a motoring journalist who's earned the nickname Bullitt. On every car trip he goes on, he does exactly the same things as McQueen, only without the style, the intelligence or the ability – basically he just drives like an idiot.

I had first heard of Bullitt on a trip to the South of France to launch the original MG Maestro. This was one of the first cars to have a built-in voice synthesiser whose metallic tones would tell you if the back door had been left open, or if the handbrake was still on, that kind of thing. I found such systems disconcerting, especially if you were driving by yourself at night, all was quiet, you were cruising along, and suddenly this voice out of nowhere said, 'Rear light bulb defective.' The shock would often be enough to make you swerve off the road and up the banking (I suspect this is one of the reasons why these systems are no longer built into cars).

However, the MG Maestro did have one, and on this trip, Bullitt was driving by himself because no-one was foolhardy enough to go with him. At some stage on the twisty mountain roads of Southern France he got into a race with another journalist, who presumably had had too much protein for breakfast. He was in front of Bullitt and this was fine for a while and really good fun, he said, because Bullitt was right behind him but just couldn't get past.

The high-speed duel went on for some miles until they came to a particularly slow, twisty stretch of road. The driver in front threaded his car through the bends, and so did Bullitt, close enough behind so that our man could see the whites of his eyes. Then Bullitt misjudged it and clipped a piece of kerbing. The Maestro became ever so slightly airborne and twisted over onto its side, coming to rest upside-down by the side of the road, half in, half out of a ditch. The man in front stopped and reversed back – was Bullitt alright?

Other journalists arrived on the scene in time to see the

door to Bullitt's car being opened. He hung there in his seatbelt, upside down, quite uninjured, as the voice synthesizer kept monotonously repeating, 'Your oil pressure is low, your oil pressure is low.' To be fair, this was the least of his problems. As they released Bullitt and he thudded into the roof, the synthesizer clicked and said, 'Your seatbelt is unfastened, your seatbelt is unfastened.' At dinner that evening someone drily suggested that Bullitt's car should be equipped with a few extra messages, such as, 'You are about to crash', 'You are turning over' and 'You are a pillock'.

One of the most embarrassing things I ever heard on a press outing came on a Vauxhall trip in the early 1980s. It all started with an evening meal at a hotel near Gatwick, prior to an early flight the following morning. All of the Vauxhall top brass were at the dinner table with us scribblers, and a fine spread of food had been laid on.

The meal began with Vichyssoise, the soup made from leeks, potatoes, chicken stock and cream. As most will know, Vichyssoise is served chilled – but this was too much for one of the hacks to comprehend. You've guessed it, he shouted across the table, 'Would you darn well believe it, soup's bloody cold.'

I went on a Honda trip in the early 1980s where the new front-wheel drive Accord was launched. The public relations staff at the time were what can best be described as 'inexperienced' in organising press trips. I say this with a large dollop of kindness since in fact, the trip was a miserable experience for all of us scribblers, compounded by the fact that for some reason these PR people had decided to save as much money as they could by doing various deals with ferry companies, hotels and airlines to make the trip as inexpensive as possible.

So, for some arcane reason, we landed in Belgium and then had to board a bus to be driven just across the

MOTORING  *SHORTS*

border into France where we were to stay in a monastery which had been converted into a hotel. Problem number one was that the main PR staff were already at the hotel; number two was that the PR woman we were with had no clear idea of where the hotel was: number three, neither did the bus driver, who spoke no English. The PR woman spoke no French, and none of the journalists spoke more than a smattering of French either.

Drink had been supplied at Gatwick prior to take-off and a good many of the hacks were pretty drunk as the bus trundled on through the night. Normally when you cross the border in Europe the customs men do not bother stopping you, but this time they did and several Belgian policemen came aboard. For some of the most drunken of the journalists, this was like a declaration of war, and various verbal assaults followed, culminating in one of Fleet Street's finest grappling with one of the Belgian coppers. The police retaliated by demanding money because we were taking cameras into Belgium! Of course, I was the only one with Belgian money, so I had to pay for everyone, and I tell you, I still haven't got the money back from Honda!

Anyway, off we go, into France, which is very dark by now. We get lost. The Honda PR woman's response to this turn of events, I kid you not, is to launch into the traditional English pastime of shouting at the bus driver in a foreign-sounding accent: 'Do you know vere ze hotel eez?' He hasn't a clue what she is talking about, and understands even less when she starts doing her impression of fat Franciscan monks, except he starts to panic because he thinks she is saying she is pregnant and about to deliver, right there on his bus.

I don't know how, but eventually we got there. But surprise, surprise, there has been a mistake. There are either not enough rooms or there are too many journalists. I don't mind sharing but I would rather share with someone who doesn't snore. No chance. I have to share with a guy who immediately falls asleep and snores like I don't know what. Suffice to say, I get no sleep, and neither do the rest of the party.

The trip did not get any better, especially the part

MOTORING  *SHORTS*

where we had to drive the cars back to Britain via a free ferry crossing which Honda had managed to get on Sally Line. The longest possible crossing, of course, and unfortunately one of the roughest I've ever experienced.

Now you know why I don't go on many trips anymore.